Laboratory Introductio
to
Bio-inorganic Chemistry

Laboratory Introduction to Bio-inorganic Chemistry

Ei-Ichiro Ochiai, BSc, PhD
Department of Chemistry
University of British Columbia
and
David R. Williams, BSc, PhD, CChem, FRIC, DSc
Department of Chemistry
University of Wales Institute of Science and Technology
Cardiff

M

First published 1979 by
THE MACMILLAN PRESS LTD
London and Basingstoke
Associated companies in Delhi Dublin
Hong Kong Johannesburg Lagos Melbourne
New York Singapore and Tokyo

Typeset by
Reproduction Drawings Ltd, Sutton, Surrey

Printed in Great Britain by
Unwin Brothers Limited
The Gresham Press, Old Woking, Surrey
A member of the Staples Printing Group

British Library Cataloguing in Publication Data

Ochiai, Ei-Ichiro
Laboratory introduction to bio-inorganic chemistry.
1. Biological chemistry–Laboratory manuals
2. Chemistry, inorganic–Laboratory manuals
I. Title II. Williams, David Raymond
574.1′921′028 QP531

ISBN 0-333-23680-7
ISBN 0-333-23681-5 Pbk

Contents

Part B
Inorganic Chemistry Relevant to Biological Sciences

Part C
Enzymes, Proteins and Model Studies

Preface

Within the last 20 years there have been many advances in our understanding of the role of elements *in vivo*. Such achievements arise from painstaking researches in many disciplines–biochemistry, inorganic reaction mechanisms, medicine, pharmacology–to mention but a few. The need to supply a steady stream of researchers has been reflected in a demand for bio-inorganic courses at undergraduate level. This trend has been supplemented with a wide range of service courses for medical students and applied biologists. Several textbooks have been published and there is a clear need to reinforce lecture and textbook instruction with laboratory training and experience. This book is a laboratory manual which collects experiments that may be used to drive home the principles of bio-inorganic chemistry.

Readers who expect to find experiments in which, for example, metal salts are administered to an animal and then some sophisticated scheme of reaction mechanism is derived, will be disappointed! Many of these experiments are fundamental to inorganic chemistry or pharmacology but they have now been adapted to consider either biological material or the presence of metals. The relationships between such experiments and bio-inorganic principles are indicated in the Introductions but it must be emphasised that the activation energy necessary to turn these experiments into direct educational relevance for each specific lecture course being mounted must, and can only, come from the course instructors.

Since the human body contains about 10^6 different solutions (i.e. solutions entrapped inside living cells) we have arbitrarily classified experiments into analysis of components of a cell, the inorganic chemistry occurring *in vivo* and the metalo-enzyme-metalloprotein chemistry occurring in the micro environments of enzyme active sites within cells. There could have been many other types of classification since bio-inorganic chemistry is contingent upon so many branches of natural science.

We have procured experiments from many sources and obtained our own experiences before writing them herein. However, we should like to express our thanks particularly to Professor A. G. Splittgerber of Gustavus Adolphus College who is the original developer of Exp.18(a) and (b), Professor W. H. Sawyer of the University of Melbourne (for Exp.18(c)), Dr. Y. Koga of the University of British Columbia (for Exp.6(a)), Dr. J. T. Roos of Marlborough College (for Exp.3-7), Mr. P. M. May of UWIST 9 for Exp.11 and 25), Dr. D. E. Fenton of Sheffield University (Exp.18) and Ms K. Oliver (Exp.21). We should also like to thank Mr. B. Clifford for technical assistance in developing some of the experiments at the University of British Columbia.

We would value receiving ideas from fellow instructors with a view to an improved second edition.

'We gather a posy of other men's flowers but the thread which connects them is our own.'

British Columbia and Cardiff, 1979 E.-I.O.
D.R.W.

Part A

Analysis of Trace Elements in Biological Materials

Chapter I

Introduction: Bio-inorganic Chemistry as a Laboratory Discipline

The impact of bio-inorganic chemistry has settled solidly into undergraduate chemistry, biochemistry and medical science courses. This volume is written as a sister volume to the four main text-books for these courses (*see* Further Reading) so that the principles learned in the lecture theatre and library can be driven home by laboratory investigation.

Depending upon the level at which bio-inorganic chemistry is being taught, and, perhaps, the degree of doubt in a student's mind about a concept, there will be differing requirements for laboratory experience to reinforce such concepts. Thus, we offer a choice of experiments so that courses can be closely chosen to suit an individual's needs, and, also, the majority of the experiments are open-ended in both degree of accuracy and scope so that personal research investigations might stem out of many of the basic experiments. This personalised tailoring of courses is indispensable because of the interdisciplinary nature of the subject – clearly, there are experiments which, to an inorganic chemist, might seem trivial whereas, to a modern biochemist, they constitute a vital part of the undergraduate training and vice versa.

Volumes have already been published which discuss the basic principles of bio-inorganic chemistry and the means through which the concentrations of metals present in any site or body fluid are dependent upon such factors as concentration of anions present, solubility products, pH considerations and enrichment factors. Thus, the need to analyse precisely for metal concentrations ought to be obvious to clinical, medical and pharmacology students.

Why is man so dependent upon trace elements? All living organisms are enveloped by the inorganic environment of this planet and, during the millions of years of man's evolution, certain dependencies upon the composition of, and degree of contamination of, the hydrosphere have been established. Very recently, on the evolutionary time scale, human exploitations of nature and the explosion in technological innovations have sometimes exceeded the speed at which man can adapt to his environment and so problems of pollution have occurred. Such problems cannot be tackled until the roles of the inorganic elements essential to man have been extensively researched. This research can only start in the laboratory since theory and experiment must progress in unison.

There are, in fact, twenty six elements of prime interest, and several of almost equal importance. These fall into four broad classifications – the *essential, beneficial, contaminating* and *polluting* elements (*see Figure I.1*). These groups parallel the concentrations in the environment during our evolution (we have evolved to use the elements most readily available in the hydrosphere – these are the essential and beneficial ones listed) and nowadays the contaminating elements (these vary from person to person but each of us contain some 20–30 such elements) and the polluting elements (especially lead, cadmium and mercury) which are figments of our more recent environment.

All four groups of elements have effects upon our bodies that are concentration-dependent (*see Figure I.2*) – the optimum concentration usually varying

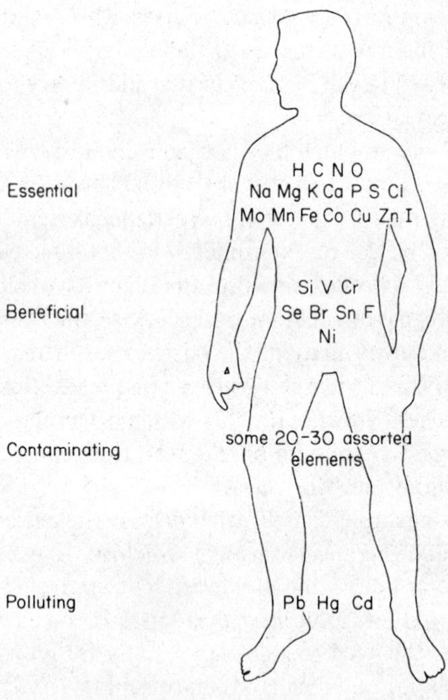

Figure I.1 Elements affecting our biochemistry

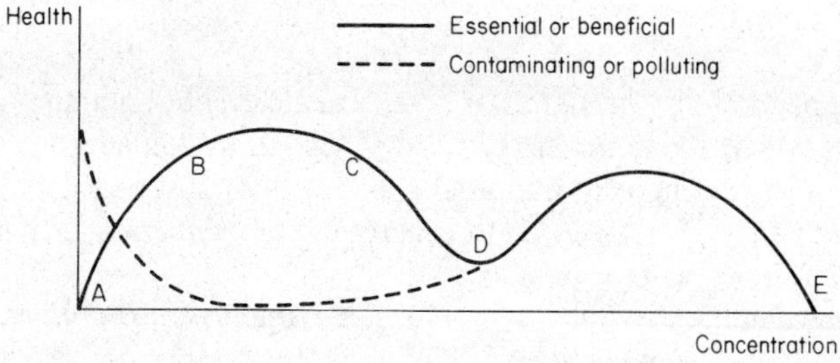

Figure I.2 The elemental concentration-dependence of health

from organ to organ and from animal species to animal species. For any given essential or beneficial element in any given body site there are clearly deficiency syndromes (A–B), optimum health concentrations (B–C) and excess or 'surplus-to-requirements' situations (C–D), i.e. the body is in negative, correct, or positive elemental balance. For contaminating or polluting elements, the concentrations normally recommended for optimum health is zero, or near to it.

The second hump in Figure I.2 (D–E) is conceptually more difficult; it depicts the administration of an element (belonging to any of the four groups) to stimulate the body's defence mechanisms to affect, say, tumour-cell reproduction (e.g. Pt/Pd compounds), wound healing (e.g. Zn compounds) or intestinal ulceration (e.g. Bi compounds). Whereas the first, optimum-health, peak refers to the lifespan of the body, the second, pharmacological, peak is one of much shorter temporal duration (a matter of days). The right-hand limit to the graph is death caused by too much of any element. Additional principles contained in such curves (*see* text-books) are the mutual interdependence of elements (e.g. Mg versus Ca) and the cyclical variation of optimum concentrations on a daily basis (the circadian rhythms of the body's chemicals).

However, we have not repeated theoretical considerations that are already readily available since it would only inflate the price of this laboratory manual. Nevertheless, we do most sincerely stress the necessity to acquaint oneself with these principles before setting out on the experiments concerned. History has taught us that scientific progress can only be guaranteed if theory and experiment journey side by side. Thus, each experiment has a brief introduction which indicates the principles about to be exercised.

Many of these experiments are exercises designed to discipline the mind to think in bio-inorganic terms rather than experiments which involve living bio-molecules—these latter would run to projects of several weeks, i.e. there is a large gap between an isolated experiment and the real multiphase, multi-component phenomena of humans. We hope that this very interdisciplinary complexity of the subject stimulates rather than deters students.

FURTHER READING (BIO-INORGANIC TEXTBOOKS)

D. R. Williams (1971). *The Metals of Life,* Van Nostrand, London

An Introduction to Bio-inorganic Chemistry, Ed. D. R. Williams, C. C. Thomas, Springfield, Illinois (1976)

E.-I. Ochiai (1977). *Bio-inorganic Chemistry, An Introduction,* Allyn and Bacon, Boston

A. M. Fiabane and D. R. Williams (1977). *The Principles of Bio-inorganic Chemistry,* The Chemical Society, London, Monographs for Teachers No. 31

Chapter II

Treatment of Biological Material

Every conceivable analytical technique has been applied to the determination of elemental concentrations of biological systems – some methods are more useful than others; the two main techniques currently in widespread use are neutron activation analysis (regrettably not a technique that can be handled by junior student bio-inorganic chemists) and atomic spectrophotometry (emission, fluorescence, and absorption). Since the distribution of elements (the ionogram) in animals and plants reflects the ionogram of the hydrosphere it is not surprising that, for practically every element being analysed, humans are capable of contaminating the sample with some additional traces of the element from their own fingers, breath, hairs, clothing, etc. This interaction between the worker and his materials is known as 'handling difficulties'. Thus, the laboratory investigation of clinical and pathological states requires reliable methods, technical skill and adequate instrumentation.

The quantitative analyses often need to be performed on relatively small samples and sometimes it is impossible to repeat an analysis on a particular specimen (this is especially so in forensic science) and so practice in sample handling on blank or non-important material is essential. Sensitivity and precision are not the sole factors: for routine analysis the overall method ought to be sufficiently 'robust' so as not to be sensitive to minor inaccuracies in burettes, wavelength settings, etc. and to changes of technical operator. It is frequently axiomatic that a method which is valuable in research can be unsuitable in routine use.

Many of the techniques employed in bio-inorganic chemistry are still the standard wet-chemistry practices of inorganic or analytical chemistry – the digestion of materials by progressively stronger acids (up to aqua regia) until completely dissolved, total combustion, titrations up to an end point. In bio-inorganic chemistry more than ever it is important to have a grasp of the theory behind the operation. The minutiae of theory, technique and methodology are given in standard inorganic practical reference books but we invite you to take any inorganic analytical experiment and attempt to perform the analysis for the same element in sea-water, blood, etc. The inherent difficulties involved in the competition by many additional elements present will soon become very obvious.

All materials, samples, equipment and reagents must be subject to strict attention in respect of clean handling or preferably no handling at all. Materials from animals can be divided into two sections. The main bulk is 'wet' whereas the surface samples of skin, hair and nail may be described as 'dry'. Wet tissue is removed from the body usually by a non-analyst and this is the time at which the contamination risk is high. These contaminants come from a variety of objects in contact with the sample – compounds adhering to the surface of the body (dirt, industrial lubricants and chemicals, cosmetics, etc.), pathologists' gloves (talcum powder, disinfectants, fillers in the rubber gloves), the instruments used (metallic dissolution or shavings) and the dissecting table (dust, disinfectants, detergents). The sample ought then to be sealed in polyethylene and not in preservatives (formalin, etc. is a rich source of contamination).

When the analyst receives the sample he endeavours to dissect out a piece of uncontaminated tissue from the centre by using silica or plastic instruments (i.e. knives which do not contain the elements sought after). At this stage it is essential not only to have a reliable method for analysis available but also to have had it pre-tested by blank runs and data for known material. It is far too late to start technique development when the sample arrives. The analysts have to be trained and the techniques have to be 'debugged' on normal, readily available samples. Some of this can be achieved by working through the ensuing experiments.

RECOMMENDED FURTHER READING

G. S. Fell and H. Smith (1976). 'General Analytical Methods', in *An Introduction to Bio-inorganic Chemistry,* Ed. D. R. Williams, C. C. Thomas, Springfield, Illinois, 254–280

T. A. Hyde and T. F. Draisey (1974). *Principles of Chemical Pathology,* Butterworths, London

A. I. Vogel (1966). *A Textbook of Quantitative Inorganic Analysis,* Longmans, London, 3rd edn.

Chapter III

Laboratory Experiments

Experiment 1
Determination of Chloride in Biological Fluids by Thermometric Titrimetry

INTRODUCTION

The determinations of chloride against silver by gravimetric or volumetric methods are well documented in textbooks of inorganic quantitative analysis. Complications that may arise in biological liquids include the interference of other ions and the selection of suitable indicators to give colour changes detectable in the crimson of blood as a solvent. Thermometric titrimetry is a means of circumnavigating some of these difficulties – the detection of end points of titrations from the inflection point in the graph of heat evolved versus amount of titrant added. *Figure III.1.1* is a thermogram of the simplest and most well-known type of titration – the titration of alkali from a burette into mineral acid in a stirred vessel ($H^+ + OH^- \rightarrow H_2O$).

The thermogram falls into three distinct regions. During a, the fore period, the vessel and contents are slowly gaining or losing heat from the surroundings and from the heat of stirring. Then the tap on the burette is opened and the titration commences.

During b, the reaction period, heat is being evolved – mainly from the heat of formation of water but possibly with a small contribution from the heats of dilution of titrant and titrate. Eventually the end point is passed and after this we are merely adding alkali to alkali and enter c, the aft period. This is sometimes called the correction period since heat changes here arise only from heats of dilution and from the fact that the titrant may not be at exactly the same temperature as the titrate.

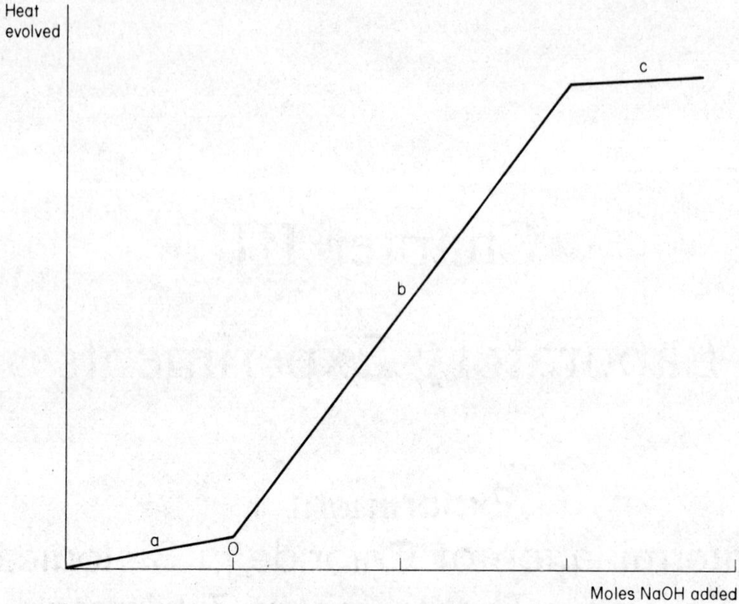

Figure III.1.1 Thermogram for the titration of sodium hydroxide into hydrochloric acid. The inflection point is the end point of the titration

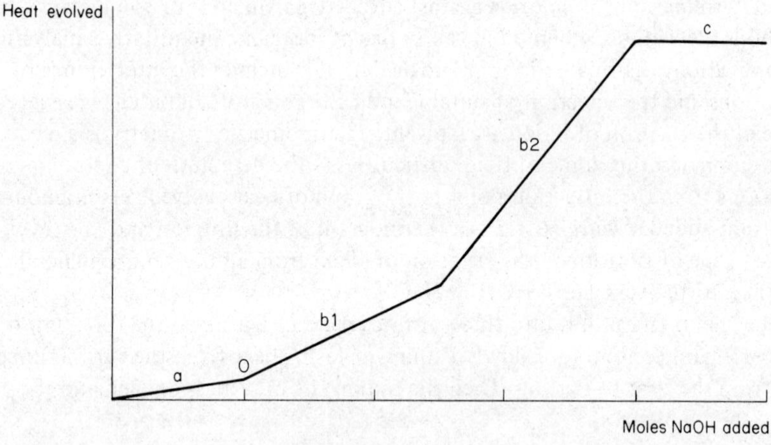

Figure III.1.2 Thermogram for the titration of HCl/$CH_2ClCOOH$ solution with NaOH. Region b1 corresponds to the HCl titre and b2 to the $CH_2ClCOOH$ titre

Solutions containing two different acids can be titrated in the same way and the thermogram has two inflection points (*Figure III.1.2*), and ligands can be titrated with metal ions to determine the end points from complex formation (*Figure III.1.3*). Coloured indicators to detect some of these end points are not easy to find.

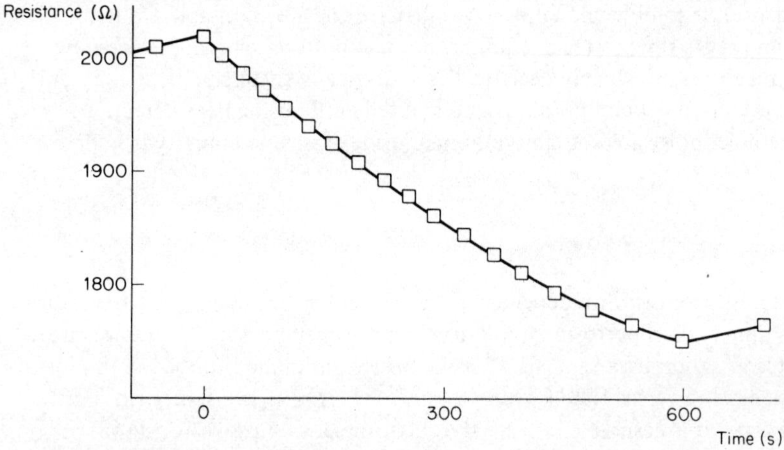

Figure III.1.3 Thermometric titration of chloride (1 mol dm^{-3} NaCl) into a mercuric (500 mmol dm^{-3} $Hg(NO_3)_2$, acidified) solution. The inflection point (at 9 min 55 s) occurs when $HgCl_2$ formation is complete

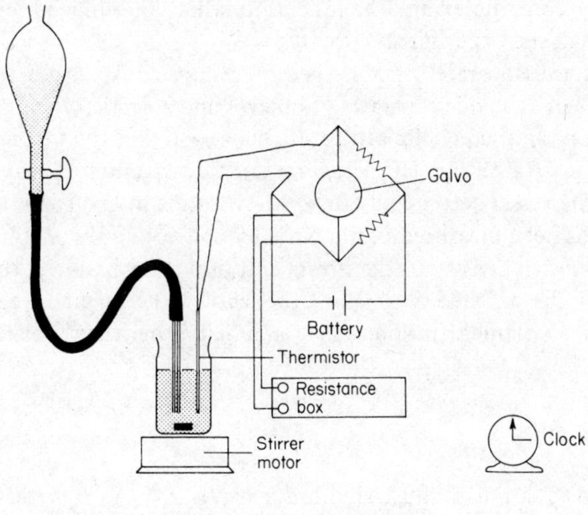

Figure III.1.4 The thermometric titration apparatus including the Wheatstone bridge circuit

SUITABLE REACTIONS

The book by Tyrrell and Beezer (*see* References) lists more than 100 aqueous reactions that have been studied by thermometric titrimetry and many more for non-aqueous solvents. In general, any solution reaction can be studied in the apparatus shown in *Figure III.1.4* as long as it goes reasonably well to completion, is soluble enough to give sufficient heat effects and is not a reaction that is slow to reach equilibrium. It is quite fascinating to see how many reactions, previously described as analytically useless because they did not give a sharp pH end point, can be titrated easily using heat detection; for example, boric acid or ammonium chloride against sodium hydroxide.

PROCEDURE

The success of a titration depends on having accurate readings of temperature versus time and on performing the titration at constant rate. The temperature detector is a thermistor and Wheatstone bridge circuit and the required plot of temperature versus time is really a plot of thermistor resistance (in ohms) versus time. It is easier to measure the time for set thermistor resistances rather than the resistance at set times. Thus the approach is to dial up a resistance value, say 2345.6 Ω, and note the time on the stopclock when the galvanometer indicates null deflection, then change the resistance to, say 2345.2 Ω and note its corresponding time, and so on. A resistance versus time plot is constructed and the inflection point determined by drawing lines through the experimental points.

A constant flow rate is maintained by using a constant pressure head device – such as a wide separating funnel – and a capillary and screw-clip. The capillary and screw-clip setting are chosen so that the theoretical end point is reached in 10–15 min. The actual rate of flow per minute may be found by weighing the vessel before and after a known time interval and, if the titrate solution has been measured out in cm^3, by converting the weight to a volume using the known density of the titrant solution. Alternatively, the titrate solution can be weighed out making the whole experiment a weight titration rather than a volume titration – that is, molal rather than molar.

APPARATUS

The simple apparatus is illustrated in *Figure III.1.4*. A magnetic stirrer and follower stirs the solution contained in the inner beaker, which is shielded from draughts by an outer beaker. A Perspex or polystyrene cap covers these beakers and supports the thermistor (Stantel F 23) and 'burette' capillary tip.

The thermistor is connected into one arm of the Wheatstone bridge and the resistance box into the facing arm. The remaining arms are two 2-kΩ resistances (± 2%, available from any radio dealer). Electromotive force (emf) is supplied from a U2 (1.5 V) battery; this gives about 1 mA in the circuit, ideal for the thermistor. A galvanometer or null point detector is connected across the bridge. The apparatus functions best when the following points are incorporated. It is very important to ensure that titrant and titrate solutions are at the same temperature and it is prudent to make them up the previous day and leave them beside the apparatus overnight.

The rate of flow is best determined *in situ*. If one tries to measure this into a clean beaker, possibly at a different level and without stirring, spurious results can arise. Stirring possibly aids the flow of titrant out of the capillary and therefore affects the flow rate.

It is important to have both the separating funnel tap and the screw-clip. The former is used merely as an on–off switch, the latter as a flow regulator. To attempt to control both functions by either the tap or clip alone is time-consuming and tends to give an uneven flow rate.

The temperature change is dependent on the number of moles of reactants and products and the volume of liquid and glass present, i.e. the higher the concentrations the greater the temperature change (but also the greater the heats of dilution). The thermistor quoted changes by about $-60\ \Omega\ K^{-1}$ at room temperature. Note that its resistance decreases as temperature rises.

SUGGESTIONS FOR PROJECTS

Set up the equipment and ensure that you can use it for aqueous titrations. Then progress to urine, saliva, stomach fluid, blood serum, etc. Investigate the possible effects of interfering ions (e.g. bromide and iodide versus chloride) and suggest means of improving the apparatus (e.g. peristaltic pumps?, thermostat baths?).

REFERENCES

D. R. Williams (1971). 'Thermometric Titrimetry'. *Educ. Chem.*, 8, 97

H. J. V. Tyrrell and A. E. Beezer (1968). *Thermometric Titrimetry*, Chapman and Hall, London

Experiment 2
A Copper(II) Ion Sensitive Electrode for Copper Assay

INTRODUCTION

Animals and plants alike are dependent upon controlled concentrations of copper ion. Imbalances in these concentrations, be they from pesticides decomposing in soil or from contaminated food and drink being administered to animals, can produce poisoning. Conversely, periodic monitoring of copper concentrations in man can be used as a physiological health indicator; an average adult contains 100 mg and an average daily diet contains 2-5 mg of which 1 mg is absorbed and appears in the blood plasma (98% of plasma copper is in the form of ceruloplasmin).

Surplus ceruloplasmin copper is excreted into the intestine via the bile duct and so ^{64}Cu studies have shown that 80% of ingested copper, under conditions of health, eventually appears in the stools, less than 1% in the urine, and the computed concentration of free copper ions in plasma is 10^{-11} mol dm^{-3}.

The copper contents of *in vivo* fluids and foods require periodic monitoring because of the following considerations. (1) Wilson has reported a genetic disease (hepatolenticular degeneration) accompanying a ceruloplasmin deficiency which produces an increased free copper ion concentration in blood plasma, a positive copper balance in tissues and an increased renal excretion. D-Penicillamine is used as a palliative and this results in increased urinary excretion of copper. (2) Blood copper concentrations increase 2-300% during infectious diseases, rheumatoid arthritis, and cancer. (3) Standard limits have been laid down for the copper contents of foods – the recommended being 7 ppm and the statutory being 20-30 ppm in the United Kingdom and cases of poisoning that have exceeded these figures are well known.

PROCEDURE

The following electrode system is required:

saturated sodium chloride calomel reference electrode	equilibrium solution	glass electrode, copper amalgam electrode

These electrodes are placed in the glass vessel shown in *Figure III.2.1*. Under ideal thermodynamic conditions at 25.00°C, the emfs (in mV) of the glass and amalgam electrode system are respectively

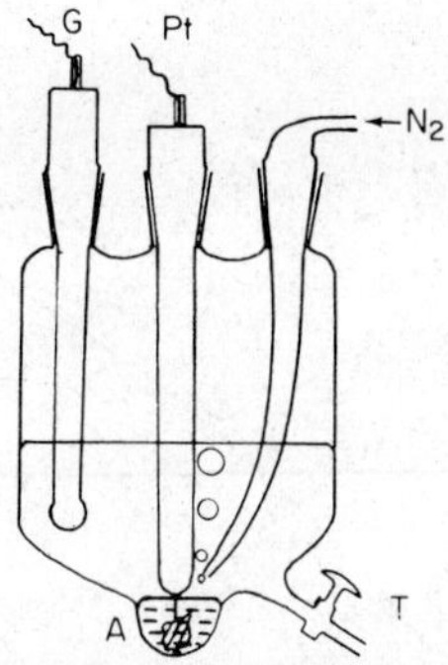

Figure III.2.1 Vessel and electrodes for amalgam analysis of copper solutions. A is the amalgam pool containing a platinum electrode (Pt). The solution pH is measured using a glass electrode (G) and this solution may then be drained away through tap T. Additional openings in the vessel top (not shown) are for the nitrogen outlet, the calomel reference electrode and for inserting the solution under test

$$E_g = E_{0g} - 59.155 \log[H^+]\cdot f_{H^+}$$

and

$$E_a = E_{0a} - 29.577 \log[Cu^{2+}]\cdot f_{Cu^{2+}}$$

Amalgam electrode complications can arise because of hydroxide formation so this is minimised and held constant by working at a constant acid pH. We suggest pH 3.0.

The experimental procedure is as follows. Standard solutions of copper are prepared (2×10^{-x} mol dm^{-3} where $x = 2$-9). The sample to be determined is digested with twice its volume of concentrated nitric acid and concentrated sulphuric acid (5 ml) by cautious heating. Further oxidising acids are added until SO_3 fumes are evolved, and all the organic matter has been destroyed. Nitrosyl sulphuric acid is removed by cooling, adding water, filtering off insoluble matter and making the neutral solution up to 100 ml. For highest accuracy it is recommended that the volume contains 25 μg of copper.

An aliquot of this solution (25 ml) is added to $HClO_4$ (25 ml, 2×10^{-3} mol dm^{-3}) in the vessel and E_g and E_a are noted. The vessel is washed out and standard solutions on either side of E_a are inserted in turn (25 ml of each) and to each $HClO_4$ is added (25 ml, 2×10^{-3} mol dm^{-3}). E_g and E_a are noted and a graph of E_a versus $\log[Cu^{2+}]$ is plotted; thus the unknown copper concentration may be read off (*see Figure III.2.2*). Such graphs do not have the theoretical slope of 29.577 mV per $\log[Cu^{2+}]$ for several reasons: there is a variation in fugacities, temperature, and chloride concentrations (from calomel leakage), and the amalgam sometimes oxidises or accepts some free Cu^{2+} causing the emf to level off slightly at $-\log[Cu^{2+}] = 6$ and above. (Nevertheless, these deviations from ideality are reproducible and we have found that our analysis of unknown copper solutions in the range 1×10^{-6} -5×10^{-4} mol dm^{-3} agreed to within 3% with analyses by electrodeposition.)

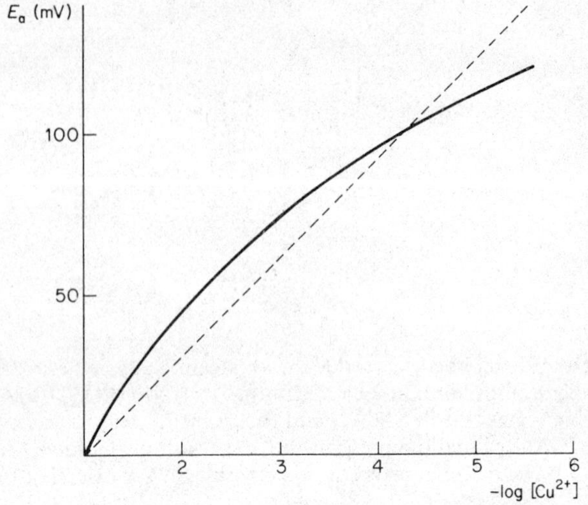

Figure III.2.2 A typical calibration curve (full line). The broken line is theoretical calibration curve of 29.577 mV$(\log[Cu^{2+}])^{-1}$ at 25°C

SUGGESTIONS FOR PROJECTS

Apply the method to the copper assay of mains tap water, soft drinks, urine or blood and compare your results with the values given in *Table III.2.1.* If there are discrepancies suggest sources of likely experimental error or reasons why, for example, the copper content of mains drinking water might be particularly high.

Table III.2.1. Legal and Measured Copper Concentrations. Units = μmol dm^{-3}

Food	
Recommended limit	110
Statutory limit	315
Average diet	15–30
Mains tap water	0–30
Average content of whole body	20
Soft drinks, uncontaminated	~30
Soft drinks, contaminated by pourer as described in first reference	1000–3500
Blood	
Healthy average, total	14–21
Healthy average, free ions	0.00001
Wilson's disease	5–16
Infectious diseases	28–42
Pregnancy	32–55
Urine	
Healthy average	0.1–0.9
Wilson's disease	2 -14

REFERENCES

I. A. Matheson and D. R. Williams (1973). 'A Potentiometric Cu Assay in Normal and Copper-Poisoned Humans'. *J. Chem. Educ.*, **50**, 345

J. Peisach, P. Aisen and W. E. Blumberg (1966). *The Biochemistry of Copper,* Academic Press, New York

G. J. Moody and J. D. R. Thomas (1976). 'Selective Ion Sensitive Electrodes', in *An Introduction to Bio-inorganic Chemistry,* Ed. D. R. Williams, C. C. Thomas, Springfield, Illinois, 220-235

Experiment 3
Atomic Absorption Analysis of Manganese and Zinc in Leaves

INTRODUCTION

Atoms need to absorb light in order to emit it, and this absorption occurs at wavelengths which are characteristic of the atom. If cations are sprayed into a hot flame, their presence and concentration may be detected by measuring the extent to which they absorb a light passed through the flame. Atomic absorption spectrophotometers are usually composed of a light source such as a hollow cathode lamp, a small quantity of the element under examination being within the cathode so that only light of the characteristic wavelength is emitted; this light is shone through a long, narrow flame into which the sample is sprayed, thence via a monochromator to a photodetector. The proportion of light absorbed follows the Beer–Lambert law, i.e. it is directly proportional to the number of absorbing ions present in the flame. The student ought to know the characteristics, sensitivity and limitations of the instrument before attempting to obtain reliable data involving an unknown sample. The following four test procedures are suggested.

TEST PROCEDURE NUMBER 1: THE EFFECT OF SLITWIDTH

Stock solutions of cobalt, nickel and iron are required (each of 1000 ppm) (ppm = parts per million). Prepare a series of calibration standards (0, 10, 20, 40, 60 ppm) for *one* of the three metals listed above as follows: into five 100-ml flasks transfer from a burette 0, 1.0, 2.0, 4.0, 6.0 ml of the metal ion stock solution. Add 1 ml of dilute hydrochloric acid to each flask, and dilute to the mark with deionised water.

Using the manufacturer's recommended instrumental settings for the metal in question, vary the slitwidth from 0.05 to 0.4 mm (50–400 μm). Aspirate the prepared standard solutions successively at each slitwidth setting, and measure the absorbance in each case.

Now set the slitwidth at 0.08 mm, and repeat the above measurements while varying the lamp current from 4 to 15 mA.

Plot calibration graphs of absorbance versus concentration (on the same graph sheet) for each value of the slitwidth/lamp current used.

Try to offer explanations for the effects which you observe. Which of the two parameters – slitwidth or lamp current – has the greatest influence on the linearity of the working curves?

TEST PROCEDURE NUMBER 2: THE EFFECT OF OBSERVATION HEIGHT ON SENSITIVITY

Stock solutions of calcium, manganese, and chromium (1000 ppm) are required.

Prepare a solution of *one* of the above metals, containing about 15 ppm of the metal concerned, by diluting the stock solution with distilled water.

Set the required wavelength on the atomic absorption instrument, adjust the slitwidth to about 0.1 mm (100 μm) and light the flame. Select the recommended acetylene flow rate. Obtain absorbance readings for your prepared solution for various values of the observation height (0.3–1.5 cm in steps of 0.2 cm).

Plot a graph of absorbance versus observation height, and offer reasons for the shape of the graph you obtain.

If desired, the investigation may be repeated using a different acetylene flow rate (either richer or leaner) in order to see the effect of varying the observation height for different fuel flow rates.

TEST PROCEDURE NUMBER 3: THE INFLUENCE OF CHANGING THE FUEL FLOW RATE ON ABSORPTION

Prepare a 15-ppm solution of one of the metals as for test number 2. Set the observation height on about 0.8–1.0 cm, and set the required wavelength for the metal in question.

Obtain absorbance readings over a wide range of fuel flow rates from a lean (non-luminous) to a fuel-rich (luminous) flame. Plot a graph of absorbance versus fuel flow rate.

Draw conclusions from your observations.

TEST PROCEDURE NUMBER 4: THE EFFECT OF CHANGING BURNER ANGLE AND/OR ABSORPTION WAVELENGTH

The analyst often wishes to determine higher concentrations of metal ions in solution than may conveniently be done under 'normal' atomic absorption conditions. One way of solving this problem is by diluting the sample solution(s) by a factor of, say, ten. However, equally satisfactory results may frequently be obtained by altering the sensitivity of the technique *either* by working at a less-sensitive wavelength for the element concerned, *or* by reducing the light path through the flame.

Stock solutions of calcium and nickel (1000 ppm) are required.

Prepare calibration standards for calcium (5, 10, 20, 30, 50, 100, 200, 300 ppm) or nickel (10, 20, 30, 50, 100, 200, 300, 500 ppm). Set the slitwidth to 0.1 mm (100 μm), adjust the observation height to about 1 cm, and set the wavelength to 422.7 nm (for calcium) or 232.0 nm (for nickel). Aspirate the standard solutions into an air–acetylene flame, starting with the most dilute standard and working up to an absorbance of 0.5–0.6.

Now rotate the burner through an angle of about 45°, and obtain a second set of readings for the calibration standards as before until once again an absorbance of about 0.6 is reached.

Re-adjust the burner parallel with the optical axis (light path), change the wavelength to 393.0 nm (for calcium) or 341.5 nm (for nickel), and repeat the readings for the standard solutions as was done previously.

Plot the three calibration curves on the same sheet of graph paper, using the same axes of absorbance and concentration in each case. What do the graphs reveal about (i) the sensitivity, and (ii) the curvature of calibration, under the three conditions employed?

We can now turn to an experiment involving 'dry' biological material (a leaf – this is easier to acquire than nail or skin) and a 'wet' fluid (fruit juice) in experiment number 4.

PROCEDURE FOR THE DETERMINATION OF MANGANESE AND ZINC IN LEAF MATERIAL

Concentrated nitric acid, concentrated perchloric acid, and stock solutions of manganese and zinc (1000 ppm) are required.

Standards

Pipette 5.0 ml of each stock solution into a 200-ml volumetric flask, and make up to volume with distilled water. Into six 100-ml volumetric flasks burette 0, 2.0, 4.0, 10.0, 15.0, 20.0 ml of Mn/Zn standard solution, add 2 ml of perchloric acid, and dilute to the mark with water. These solutions contain 0.0, 0.50, 1.00, 2.50, 3.75, 5.00 ppm of manganese and zinc, and correspond to 0, 25, 50, 125, 187.5, 250 ppm in the original sample.

Procedure

In duplicate, weigh out accurately about 2 g of the sample into a dry 250-ml conical flask. Carefully add 10 ml of concentrated nitric acid followed by 2 ml of perchloric acid; swirl to mix, place a small funnel in the neck of the flask, and allow the initial vigorous reaction to subside. Then heat the flask on a hot-plate and allow the solution to boil gently until all the nitric acid has been driven off. Should the mixture darken at this stage, add a further 5 ml of nitric acid and proceed as before. Continue in this way until, after the final traces of nitric acid have been driven off, a white or colourless solution remains. Cool the flask, dilute carefully to about 20 ml with water, filter (if necessary) through a dry Whatman 40 filter paper, and dilute to 100 ml in a volumetric flask.

Analysis

Aspirate the standards followed by the sample solutions using scale expansion if required. Plot a graph of absorbance against ppm Mn/Zn in the original sample, and read off the concentrations in the unknown sample.

Experiment 4
The Determination of Calcium and Magnesium in Fruit Juice

Stock solutions of calcium and magnesium (1000 ppm), and a lanthanum solution (5% weight/volume) are required.

STANDARDS

Dilute 50 ml of the calcium solution and 5.0 ml of the magnesium solution to 500 ml in a volumetric flask. Into five 100-ml volumetric flasks burette 0, 5.0, 10.0, 15.0, 20.0 ml of the mixed calcium/magnesium stock solution. Add 10 ml of lanthanum solution to each flask, and dilute to volume with distilled water. These solutions contain 0, 5.0, 10.0, 15.0, 20.0 ppm of calcium and 0, 0.5, 1.0, 1.5, 2.0 ppm of magnesium in 0.5% lanthanum solution, and correspond to

0, 50, 100, 150, 200 ppm of calcium and 0, 100, 200, 300, 400 ppm of magnesium in the original sample.

PROCEDURE

Take a sample of pure orange juice which has been centrifuged to remove suspended matter. Pipette 10 ml of the sample into a 100-ml volumetric flask, add 10 ml of lanthanum solution, and make up to the mark with water (Solution A). Use this solution for the determination of calcium. Dilute 5 ml of lanthanum solution before making up to the mark (Solution B). Use Solution B for the determination of magnesium.

ANALYSIS

Aspirate the blank and standard solutions followed by the samples. Plot a graph of absorbance against percentage calcium and magnesium in the original samples, and read off the concentrations in the unknown sample.

REFERENCE

G. S. Fell and H. Smith (1976). 'General Analytical Methods', in *An Introduction to Bio-inorganic Chemistry,* Ed. D. R. Williams, C. C. Thomas, Springfield, Illinois, 254–280

Experiment 5
Trace Elements in Sea Water

INTRODUCTION

The range of concentrations of trace elements normally encountered in the oceans is known to be 0.2–20 ppb (ppb = parts per 10^9) for Zn, 0.07– 10 ppb for Cu, and 1.0–20 ppb for Ni. At these low concentrations few, if any, analytical methods have sufficient sensitivity to permit direct determination. Even for the powerful atomic absorption spectrophotometry, these concentrations are comparable to the detection limits. Therefore, a concentration step is normally necessary in the analytical procedure. In this experiment, an extraction with an organic solvent serves this purpose, and a concentration increase up to 50 times is attained by the extraction.

PROCEDURE

To prepare the Solutions I–IV (with ingredients 1–6 of *Table III.5.1*), add each ingredient to the separating funnel (250-ml funnel for Solutions I–III, 500-ml funnel for Solution IV of *Table III.5.1*). Shake each solution vigorously for 3 min, then leave it for 30–60 min to allow the phases to separate. Run off the lower aqueous layer into a clean Erlenmeyer flask and keep the solution for re-extraction. Run off the organic layer into a clean 10-ml volumetric flask and make up to the mark with MIBK.

Table III.5.1 Sample Solutions (volumes in ml)

	Solution numbers			
Ingredient	I	II	III	IV
For the first extraction				
1. Distilled water	100	100	100	500 (sea water)
2. 500 mmol dm^{-3} HCl	1	1	1	5
3. Cu 1-ppm solution	3	6	10	0
4. Zn 1-ppm solution	3	6	10	0
5. APDC solution[a]	10	10	10	10
6. MIBK[b]	12	12	12	16
For the second extraction				
2. APDC solution[a]	5	5	5	5
3. MIBK[b]	10	10	10	10

[a]APDC = ammonium pyrrolidine dithiocarbamate, chelating agent.
[b]MIBK = methyl isobutyl ketone, an extracting solvent.

Transfer the solution to a clean weighing bottle. Return the once-extracted aqueous solutions to the separating funnels and again add 5 ml of APDC and 10 ml of MIBK. Do the extraction as above. The MIBK solutions thus obtained are subjected to analysis either by atomic absorption or by visible spectrophotometry. Thus either analyse for copper and zinc concentrations using the atomic absorption spectrophotometric technique that you learned in Experiments 3 and 4, or use a visible spectrophotometer (such as the Unicam SP90) and plot absorbance (the sum of those for the two extractions) against the concentration of the standard MIBK solutions (extracts from I–III) and obtain the calibration curve for each element. This curve ought to be fairly straight and it is used to interpret the absorption figure for the MIBK solution obtained from 500 ml of sea water to calculate the concentrations of copper and zinc, both in g dm^{-3}, assuming the density of sea water to be 1.032 kg dm^{-3} in ppb.

Experiment 6
Trace Elements in Hair and Blood

ZINC CONTENT OF HAIR

Introduction

A number of enzymes depend on zinc for their action, e.g. carbonic anhydrase, alcohol and lactic dehydrogenases and various peptidases. Studies have shown that zinc depletion increases lethargy and slows learning rates in animals. Zinc deficiency in the rat is manifested by retarded growth, loss of hair and lesions of the skin.

Since zinc is widely distributed in foods it is unlikely that zinc deficiency will occur in human beings eating an otherwise adequate diet. However, there is a report where a team of university wrestlers were found to have low zinc levels not only in their hair, but also in their red blood cells and urine. A group of Iranian dwarfs have been shown to be zinc deficient.

Hair consists mainly of the protein keratin which contains as much as 14% sulphur. However, a number of elements can be detected in trace amounts of hair: Mg, Al, Cl, Ca, Cr, Mn, Fe, Co, Cu, Zn, As, Se, Cd, I, Hg, Au. The amount of the trace element present in hair can often be a measure of the amount of the element ingested for a period during which growth of the hair took place. This fact has been used to prove arsenic poisoning even many years after the actual time of death.

Contamination of hair from external sources is known to take place for some elements. This, however, has not been found to be the case for zinc, even for those people using commercial shampoos containing zinc pyrithione. Zinc is present in hair in such concentrations as to be suitable for determination by atomic absorption spectroscopy. The concentration for a particular individual has also been found to be fairly constant along the length of hair and also from the position on the head from which the sample is taken.

Although there is some discrepancy in the earlier literature concerning the normal range of the zinc concentration in human hair, it is now well established as being between 150 and 200 $\mu g\ g^{-1}$. Values greater than 250 $\mu g\ g^{-1}$ are, however, not uncommon.

Procedure

Obtain a hair sample of about 1 g. Record age and sex of the hair sample donor, and also categorise the colour of hair in one of the following (black, brown, red, blond). Wash hair sample in detergent followed by thorough rinsing with water.

Rinse in methanol several times and dry in an oven at about 100°C for half an hour. Weigh out accurately two samples of hair of about 0.3 g into two separate 100-ml Erlenmeyer flasks and add carefully 10 ml of concentrated nitric acid. Digest the hair sample by gently boiling it in a fume hood. When the volume is reduced to one-half the original, cool the solution and carefully add 2 ml of $HClO_4$ and keep boiling until the volume has decreased to 1-2 ml. After cooling, the solution is made up to 100 ml in a volumetric flask. Meanwhile, prepare the standard Zn solutions with concentrations of 0.250, 0.500, 0.750, 1.00, 1.50, 2.00 ppm by diluting the stock solutions with specially prepared distilled water. Measure and record the atomic absorbances of the standard solutions and sample solutions (as for Experiments 3 and 4). Draw the calibration curve and determine the concentration of Zn in the sample solutions and hence calculate the Zn content in hair.

IRON IN WHOLE BLOOD

Procedure

(i) Prepare 100 ml of standard iron solutions of concentrations 2.0, 4.0, 6.0, 8.0 and 10.0 mg dm^{-3} (ppm), using 100-ml volumetric flasks and a stock standard iron solution (1000 mg dm^{-3}).
(ii) Prepare the sample solution by pipetting 1.98 ml of 0.2 mol dm^{-3} HCl solution into a clean dry container. Add to it 0.02 ml of the whole blood sample using a blood pipette. Cover the container with Parafilm and shake it.
(iii) Carry out the analysis by reading the absorbances of the standard solutions and the sample solutions with the atomic absorption spectrometer and determine the iron concentration in units of mg dm^{-3}, i.e. ppm.

REFERENCES

R. K. Pomeroy, N. Drikitis and Y. Koga (1975). *J. Chem Educ.*, **52**, 544
Trace Elements in Human Health and Disease – Zinc and Copper, Eds. A. S. Prasad and D. Oberleas, Academic Press, New York (1976)

Experiment 7
Estimation of Lithium in Drinking Water

INTRODUCTION

Lithium is one of the most ubiquitous elements in nature and the incidence of manic-depressive psychosis is inversely proportional to the lithium contents of drinking water. Thus plasma and water lithium analyses are frequently required. Lithium is not an essential element whereas sodium and potassium are and so all samples for analysis contain high concentrations of sodium and potassium. Rather than separate out the small amount of lithium from this overwhelming amount of ions of a similar chemistry, emission flame photometry using a lithium-specific filter is the method most widely used.

When ions of calcium and the alkali metals are heated in a flame, energy is absorbed to promote some of the orbital electrons to a higher energy level. When these ions revert to their original energy levels, light of a characteristic frequency is emitted. Analogous to atomic absorption spectrophotometry, the amount of light emitted is proportional to the number of ions present and, if this emitted light is passed through a filter and recorded photometrically, the method can be made quantitative.

A gas is pumped through the solution to be analysed and this sprays the sample into the flame (nebulising is often used to maintain spray uniformity). The emission intensity is recorded initially for a calibration curve involving different concentrations of lithium and then for the samples of drinking water, serum, etc.

PROCEDURE

Set up the flame photometer instrument in accordance with the operating instructions. Prepare a suitable calibration curve using lithium carbonate solution (full scale deflection is usually of the order of 15 ppm but check with the operating instructions). Now measure the lithium concentration of the sample of drinking water. If the sample's lithium content is too low to fall on the previously established calibration curve, concentrate the water by evaporation (rotatory evaporator). It is essential to perform a blank using deionised water by passing it through all the pre-concentration manipulations employed. For example, evaporation may well dissolve lithium ions from the surface of the glassware. Also, when taking water samples from taps, run off some water, that may have been standing in the narrow metal or plastic plumbing, before collecting the actual sample.

SUGGESTIONS

(a) Analyse samples of drinking water from several areas of your town and also from the reservoirs. Try to explain any discrepancies between samples from different sources. (Hint: how does one explain the hardness of water varying from reservoir to reservoir?)

(b) Write a short essay on 'lithium and manic-depressive psychosis'. Warning: lithium salts ought only to be prescribed by a Doctor of Medicine. Self-medication can be extremely dangerous.

REFERENCES

T. A. Hyde and T. F. Draisey (1974). *Principles of Chemical Pathology,* Butterworths, London

M. T. Doig, M. G. Heyl and D. F. Martin (1973). 'Lithium and Mental Health'. *J. Chem. Educ.,* **50**, 343

Experiment 8
Estimation of Water Contamination

INTRODUCTION

The maintenance of a pure water supply is of paramount importance to life on this planet and many science graduates are now involved in the monitoring, controlling and prevention of water pollution or contamination. (Contamination is loosely accepted as the presence of unnecessary material in the natural system. Pollution is the presence of poisonous, dangerous compounds. Clearly, the dividing line is not a sharp one since *high* concentrations of both essential and contaminating elements are usually dangerous. We do not accept the Doomsday approach to science and to chemical industrial waste so we use the term contamination in this experiment.)

The complete analysis of a sample of water would take many months and require a group of technologists embracing analytical and inorganic chemists, physicists (for gases present and radiation), bacteriologists and virologists. Indeed the complete 'structure' of pure liquid water itself, as yet, is undecided. Obviously, this mammoth team effort would only be used should a sample of water be brought here from another planet. Instead, on a routine basis, one analyses for the features most pertinent to the local situation, e.g. for iron and aluminium, respectively, in the vicinity of steelworks and aluminium smelters. Here, we aim merely to teach the principles of contamination monitoring. In all instances the degree of contamination is compared with that of a clean sample (i.e. one containing a socially desirable level of contaminant – not an easy level to define as is witnessed by the fluoridation of public water supplies saga!). We shall aim at estimating the solid matter, oxygen gas, hardness and physical characteristics of a sample of water.

PROCEDURE

Prepare some glass sample bottles (about 250 ml) by cleaning them well in an ultrasonic cleaning bath or in chromic acid. Rinse several times in distilled or deionised water and stopper with clean glass or plastic stoppers. Immerse the sealed bottles under the surface of the reservoir, pool or river under study and open the stoppers (take care not to disturb the sediment at the bottom of the pool, river, etc.). Re-stopper under water ensuring that there is no air bubble. Label the bottles with date, source, etc. and store in the refrigerator (this holds the dissolved gases in the sample). Repeat for a source of clean sample.

Physical examination

Compare samples, at the same temperature, for discolouration, dissolved solids (using Brownian scattering effect if necessary), the presence of organisms (does a mould appear after standing for several days?) – use a microscope if available–, odour and taste (carefully, and spitting out of mouth afterwards – like connoisseurs of wine!).

Material contents, soluble and suspended

The total matter present is measured by simply evaporating to dryness a 10- or 15-ml sample in a clean silica crucible with cap. The suspended matter down to a specified size can be estimated by shaking the sample and then filtering through a Millipore filter (pore sizes available are 0.01-8.0 μm) and then determination of total material by crucible heating as before. The suspended matter content may then be calculated by the difference in weights. Repeat for the clean sample.

Oxygen content

Contaminants can use up the dissolved oxygen to a level that is dangerously low for aquatic life. The quantity of oxygen present may be determined by titration against iodine/sodium thiosulphate using the D'Auria *et al.* recipe. Remove 2 ml of sample from the full bottle and replace this volume with 1 ml of manganous sulphate (3.2 mol dm^{-3}) and 1 ml KOH/KI solution (about 5:1 ratio of concentrations). Replace the stopper and shake to mix. Remove a further 1 ml and replace with 1 ml of sulphuric acid (concentrated, CARE!). Stand for 10 min or longer until the precipitate dissolves. Titrate the iodine liberated against sodium thiosulphate solution (0.0025 mol dm^{-3}). Use 100 ml of your sample and use starch as indicator. The end point corresponds to the titre at which the blue colour just disappears. The amount of dissolved oxygen in ppm = 0.2 x number of ml of thiosulphate at the end point. Repeat using a clean sample.

Hardness of water sample

Calcium and magnesium ions give water its familiar hardness. This Mg^{2+} + Ca^{2+} total concentration is measured by titration of a sample versus EDTA (about 1 mol dm^{-3}) using eriochrome black T as indicator in the presence of a pH 10 ammoniacal buffer solution until the indicator changes colour from red to blue. (A trace of potassium cyanide may be added to the titration to mask heavy metals that interfere with the indicator.) A second sample of water under examination may now be titrated against EDTA for its calcium content alone: sodium hydroxide is added to raise the pH to greater than 12, magnesium hydroxide is thus precipitated and then the calcium is titrated with EDTA as before except that murexide is now used as indicator at this higher pH. The magnesium content is found from the difference recorded in both titrations.

SUGGESTIONS FOR FURTHER STUDIES

Compare the results of your studies with the criteria for permissible and desirable levels of chemicals in public drinking water supplies (*Handbook of Analytical Toxicology,* Ed. I. Sunshine, Chemical Rubber Co., Cleveland, Ohio, 1969).

Critically assess (a) the precision of the methods you have used (e.g. what gases may have been driven off during your sample heating in the silica crucibles? – estimation of solid material?), and (b) the purity, or otherwise, of the drinking water supplies studied.

Select an additional contaminant for scrutiny, in particular one which has local pertinence.

How does the ambient temperature affect (a) the reservoir storage of water supplies, and (b) your laboratory samples?

REFERENCES

J. M. D'Auria, A. B. Gilchrist and J. J. Johnstone (1973). *Chemistry and the Environment – a Laboratory Experience,* Saunders
H. A. Flaschka, A. J. Barnard and P. E. Sturrock (1969). *Quantitative Analytical Chemistry, Vols. I and II,* Barnes and Noble, New York

Experiment 9
Air Pollution Studies

INTRODUCTION

The human body is much more susceptible to air-borne pollutants than to those transported by any other means. This vulnerability arises because man has evolved as a being capable of absorbing into the bloodstream some 60–80% of the gases present in the lungs. This may be contrasted with the epidermis (0–5% absorption) and the intestine (about 10% absorption of metal ions). The supreme dangers of lung pollution may be realised when one contemplates that one can survive for days without stomach absorption (food, water, etc.) whereas a few minutes with inadequate lung absorption of oxygen can be fatal. Thus, it is not surprising that polluting gases have been called 'public enemy number one'. The main gaseous challenges to our environment are sulphur dioxide, nitric oxide, nitrogen dioxide, carbon monoxide, carbon dioxide and hydrocarbons. In addition, absorbing surfaces can become coated with particulate matter.

Although it is relatively easy to show that air has pollutants, it is exceedingly difficult to determine their concentrations.

PROCEDURE

Insoluble solids

Place a large beaker on the roof of the building for one week or 10 days. Filter off the insoluble solids through a weighed Millipore filter, dry at 105°C and re-weigh to give the amount of insoluble solid precipitated in 7–10 days. Measure the volume of the filtrate and make up to 500 ml and set aside for later. Replace the filter on the filter holder and rinse through five times with chloroform (25 ml). This removes the tarry matter. Air-dry and re-oven dry at 105°C and finally re-weigh to give an indication of the amount of tar in the sample studied.

Soluble solids

Take a 100-ml sample of the previous aqueous filtrate and carefully evaporate to dryness in a pre-weighed evaporating dish. Cool in a desiccator and re-weigh to give the concentration of soluble solids in the 7–10 days' collection.

Chloride

To a second 100-ml aliquot add nitric acid (3 ml, 6 mol dm^{-3}), ferric nitrate solution (1 ml, 40%) and silver nitrate solution (10 ml, 10 mmol dm^{-3}). Titrate against ammonium thiocyanate (10 mmol dm^{-3}) until a red colour appears. Perform a blank titration on deionised water. The chloride content can then be calculated and scaled up to the concentration in the original sample.

Sulphur dioxide

Take a bicycle or football pump and calibrate it for the volume of air displaced per stroke by blowing air into a water-filled, inverted measuring cylinder standing in a trough of water. Calculate the number of full strokes necessary to

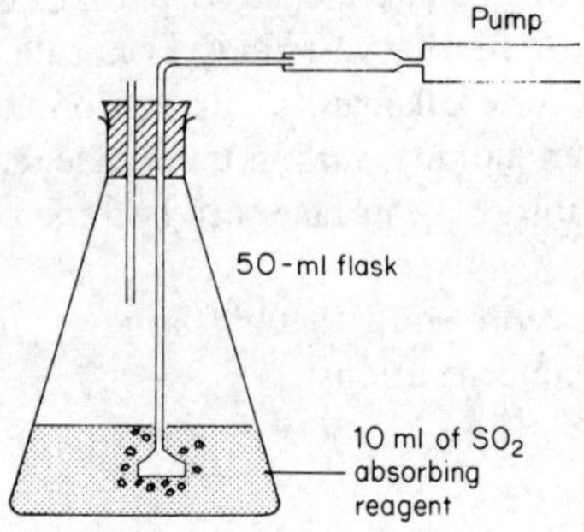

Figure III.9.1 Apparatus for SO_2 analysis

give one litre of air. Now set up the apparatus illustrated in *Figure III.9.1* and charge with sodium tetrachloromercurate solution (10 ml). Take the flask to the sampling site and pump air through the sinter at about 4 l min^{-1}. Record the volume of air passed through. Stopper the flask and return to the analytical laboratory.

This solution is now analysed by the rosaniline dye method. To the 10-ml sample add rosaniline (1.0 ml) and formaldehyde solution (1.0 ml). Now prepare a calibration curve for spectral absorption analysis. This is a series of sodium tetrachloromercurate solutions each with added rosaniline and formaldehyde as above but containing varying weighed amounts of sodium bisulphite. The absorbances of these standard solutions and of the sample are measured at 560 mμ. The blank cell contains sodium tetrachloromercurate/rosaniline/formaldehyde. The number of milligrams of SO_2 in the volume of air sampled may now be calculated.

SUGGESTIONS FOR FURTHER STUDIES

Analyse air for pollutants other than those just described.

Write an account of 'improvements made in terms of clean air during the last decade'.

REFERENCES

J. M. D'Auria, A. B. Gilchrist and J. J. Johnstone (1973). *Chemistry and the Environment – a Laboratory Experience,* Saunders

Environmental chemistry – a collection of articles appear in *J. Chem. Educ.,* **49** January 1972 issue; *Chem. Brit.,* **8** June 1972 issue

Chemistry in the Environment, Introduced by C. L. Hamilton. Readings from *Scientific American*. Freeman, San Francisco (1973)

Part B

Inorganic Chemistry Relevant to Biological Sciences

Chapter IV

The Basic Principles of Coordination Chemistry

INTRODUCTION

The manifold and painstaking researches into inorganic coordination chemistry executed over the past century by such scientists as Alfred Werner, Jannik Bjerrum, Sophus Jørgensen and Ido Leden, and by descendants of their research schools, have yielded adequate principles to found our bio-inorganic courses upon since the inception of the subject in the mid-1950s and a flourishing of interest in the early 1970s. Many of these principles are outlined in the following paragraphs, the choice being made because the principle is fundamental to coordination chemistry and/or because a subsequent experiment illustrates a concept involving the principle. The ensuing abstract is to refresh the memories of inorganic chemists and to introduce biochemistry and medical students to these principles.

When two compounds of apparently stable molecules combine to give a product that is more stable than the two reactants, the product is called a coordination compound. For example, ferrous cyanide and potassium cyanide in one-to-four ratio forms a brownish-yellow compound, potassium hexacyanoferroate:

$$Fe(CN)_2 + 4KCN \rightarrow K_4Fe(CN)_6 \qquad (IV.1)$$

The product can be regarded to be a salt composed of K^+ and $Fe(CN)_6^{4-}$ because it dissociates into these components in aqueous media and the latter behaves as an entity in solution just like the anion sulphate SO_4^{2-} or perchlorate ClO_4^-. The iron in $Fe(CN)_6^{4}$ is surrounded by six CN^- groups arranged in an octahedral manner. The iron in this entity is called the central metal ion, and the surrounding groups about it, CN^- in this case, are called 'ligands'; the whole entity $Fe(CN)_6^{4-}$ is called a 'complex ion' or 'coordination compound'. The bond between the central metal ion and a ligand is called a 'coordination bond'. It is neither a purely covalent nor a purely ionic bond; the nature of the coordination bond will be discussed later. The number of ligands directly bound to a central metal ion is called its 'coordination number'; the coordination number in $Fe(CN)_6^{4-}$ is six. The CN^- group binds with the iron through its carbon atom and forms a bond; it is thus called a 'monodentate' ligand. Other ligands can have more than two coordinating atoms. For example, the anion of the amino acid glycine, $NH_2CH_2COO^-$, binds to a metal ion, e.g. copper (II), in the following way:

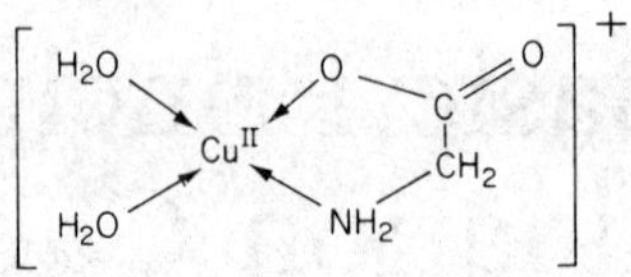

The anion binds to the metal ion through the oxygen atom of the carboxyl group and the nitrogen atom of the amino group. The ligands that bind to a metal ion by two coordination atoms are called 'bidentate'. Tridentate, tetradentate and multidentate ligands can be similarly defined. In the case of a multidentate ligand, the anion binds to a metal ion in a pincer-like fashion. Thus it is said to 'chelate' a metal ion (*chele* = claw, Greek); such a ligand is called a 'chelating' agent and the coordination compound it forms is called a 'chelate' compound. One of the most often used chelating agents is EDTA (ethylenediaminetetraacetate) and it is hexadentate and forms the chelate compound shown below.

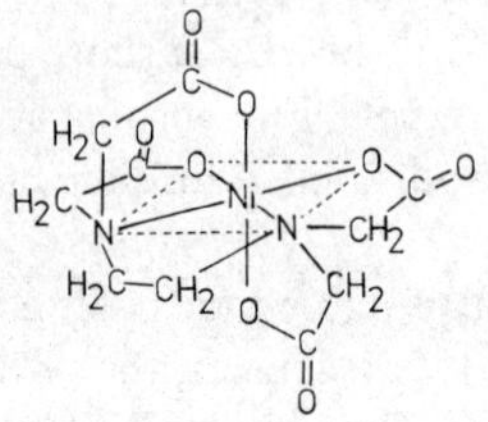

$[Ni^{II}(EDTA)]^{2-}$

Many of the compounds of bio-inorganic interest are coordination or chelate compounds. For example, the haem group (*see* Formula III below)

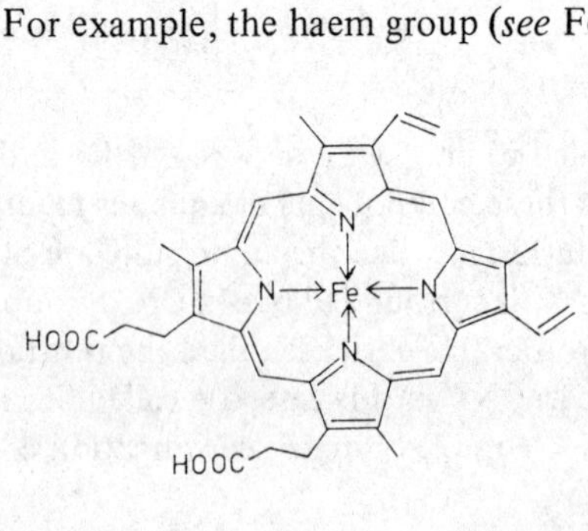

is Fe(II)- or Fe(III)-protoporphyrin-IX, and the four pyrrole nitrogen atoms of the porphyrin ring coordinate to the Fe(II) (or Fe(III)); thus it is four-coordinate and the porphyrin ring is a tetradentate ligand. In haemoglobin and myoglobin, the nitrogen atom of a histidine residue of the protein coordinates to the Fe(II), in addition to the four-coordinate porphyrin ring. The haem group in haemoglobin thus is five-coordinate. An oxygen molecule binds to the sixth coordination site of the haem group.

BONDING IN COORDINATION COMPOUNDS

The central metal ions can normally be regarded as a Lewis acid and all ligands (except for some special cases) bonded to a central metal ion have lone pair(s) of electrons. The coordination bond can be envisaged as an electron donation from the ligand to the metal ion. One extreme case of coordination bond can be almost electrostatic, as in, for example FeF_6^{3-}. Fluoride, being very electronegative, is reluctant to donate its electrons to the metal ion and the resulting bond Fe(III)-F^- is almost ionic. Another extreme of coordination bond is an almost pure covalent one. The coordination bonds usually come inbetween the purely ionic and the purely covalent one, leaning more towards the ionic bonds in most cases.

The coordination number and the shape (structure) of a metal complex are determined by many factors, the important ones being the radii of the metal ion and the ligand atoms, their radius ratio, the number of electrons on the metal ion and the type of orbitals they occupy, the so-called 'ligand field strength', and steric factors. The most common structures of metal complexes are six-coordinate octahedral, six-coordinate tetragonal, four-coordinate tetragonal (square planar) and four-coordinate tetrahedral. Typical examples are shown in *Figure IV.1*.

There are several theories to explain the nature of the coordination bond. We will briefly discuss 'ligand field' theory and 'molecular orbital' theory.

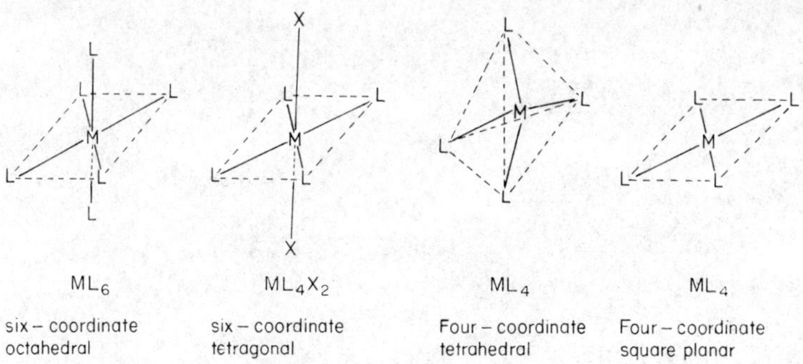

Figure IV.1 Typical structures of metal complexes

Ligand field theory

We deal here mainly with transition and post-transition metal complexes in which the outermost electrons are in *d*-type orbitals, and whose structure is octahedral. We assume that the *d*-electron system of a metal ion is placed in an electric field (called ligand field) created by the ligands arranged in an octahedral manner. In such a field the five-fold degenerate *d*-orbitals are split into two sets, e_g and t_{2g} (*see Figure IV.2*). The orbitals of the former symmet-

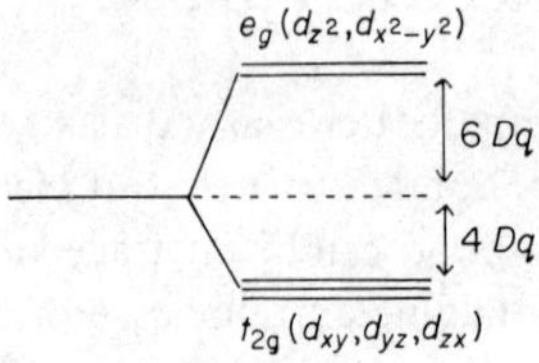

Figure IV.2 Splitting of *d*-level in an octahedral field

rical category are d_{x^2} and $d_{x^2-y^2}$ in the cartesian coordinate system; thus e_g is two-fold degenerate. The orbitals of t_{2g} symmetry are d_{xy}, d_{yz} and d_{zx}; t_{2g} is three-fold degenerate. The shapes of these orbitals are shown in *Figure IV.3.* The energy of e_g is $6Dq$ above the weighted average and that of t_{2g} is $-4Dq$; therefore, the energy difference between e_g and t_{2g} levels is $10Dq$, whose magnitude quantises the ligand field strength.

The *d* electrons occupy either e_g orbitals or t_{2g} orbitals. Let us examine an example of Fe(III); it has five *d* electrons. There are two ways to distribute these five electrons, as illustrated in *Figure IV.4.* In one case where the ligand field is rather weak, the ligands being weakly coordinating, the five electrons distribute themselves so that a maximum spin multiplicity may be attained

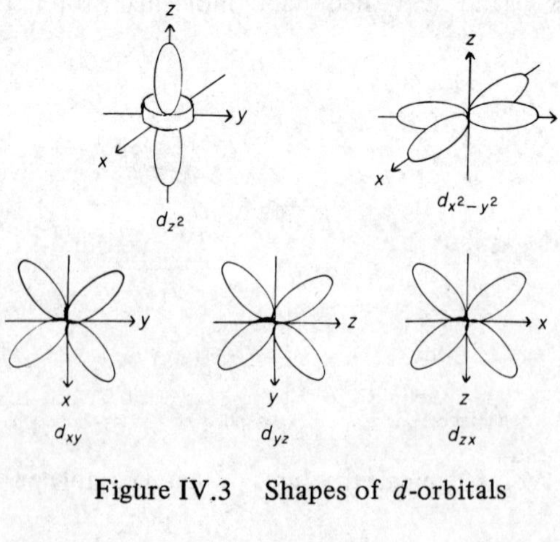

Figure IV.3 Shapes of *d*-orbitals

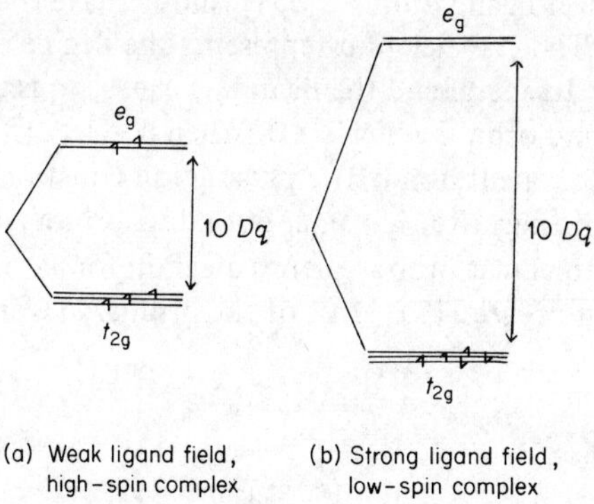

Figure IV.4 High- and low-spin complex

(Hund's rule). This leads to a high-spin complex with $S = 5/2$ in this case. On the other hand, the five electrons occupy the low-lying t_{2g} orbitals if the ligand field is strong enough, yielding a low-spin complex with $S = 1/2$. The spin multiplicity and hence the strength of ligand field and in certain cases the structure of a complex can be determined by magnetic susceptibility measurements, ESR spectroscopy, optical spectroscopy and other methods. The magnetic moment μ determined by experiments is related to the spin multiplicity by the following equation

$$\mu = g(S(S+1))^{1/2} \qquad \text{(IV.2)}$$

where g, called g-factor (or value), is 2 (more accurately 2.0023) for a free electron and can be approximated to be 2 for many metal ions of the first transition series.

A complete set of octahedral ligand field energy diagrams for all d^n (n = 2–8) can be found in the so-called Tanabe–Sugano diagram (*J. Phys. Soc. Japan*, **9**, 753 (1954)). Ligand field theory can be likewise applied to other symmetries of structure such as tetrahedron and tetragon. A ligand field treatment of a tetragonal symmetry will be found in Experiment 15.

Ligand field theory can best be applied to the interpretation of the magnetic and the optical spectroscopic properties of metal complexes, but is not very suitable in dealing with the nature of the coordination bond. The molecular orbital theory is more suitable for the latter purpose.

Molecular orbital theory

The bond nature can be treated in terms of the coefficients of a linear combination of all atomic orbitals involved in a metal complex bonding. In general, a molecular orbital can be expressed by

$$\psi_i = \alpha\phi_i \text{ (metal)} \pm (1 - \alpha^2)^{1/2}(\Sigma_j\, a_{ij}\, \chi_{ij} \text{ (ligand atom)}) \qquad \text{(IV.3)}$$

The linear combination of ligand atoms $\Sigma_j a_{ij}\chi_{ij}$ should have the same symmetry as the metal orbital ϕ_i. The coefficient α represents the degree of electron delocalisation between the metal and the ligand. If $\alpha = 1$, ψ_i is a pure metal orbital; ψ_i is a pure ligand orbital when $\alpha = 0$. When $\alpha = 1/(2)^{1/2}$ and the sign is plus, ψ_i represents a covalent bond. The plus sign in equation IV.3 represents a bonding orbital and the negative sign an antibonding orbital. The details of calculation aside, the molecular orbital energy diagram for an octahedral complex is illustrated in *Figure IV.5.* Most of the ordinary (Werner type) metal

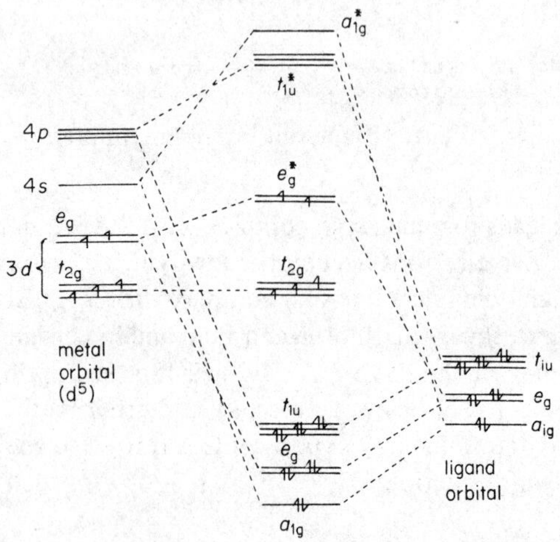

Figure IV.5 A molecular orbital energy level diagram (π-bondings are neglected)

complexes can be described by the diagram. The ligand electrons occupy the low-lying molecular orbitals of symmetry a_{1g}, e_g and t_{1u}; $(a_{1g})^2(e_g)^4(t_{1u})^6$. The symmetry a_{1g} is that of *s*-orbital orbitals, e_g is that of d_{z^2}, $d_{x^2y^2}$, and t_{1u} is that of *p*-orbital orbitals. Therefore, this configuration corresponds to the classical description of the hybrid orbital sd^2p^3. The metal electrons occupy the almost non-bonding t_{2g} orbitals and the antibonding e_g^* which is similar to the pure d_{z2} and $d_{x^2y^2}$. Thus this portion of the diagram corresponds nicely to the ligand field theoretical description. The low-lying a_{1g}, e_g and t_{1u} orbitals have their value of α (equation IV.3) closer to zero in many ordinary metal complexes. These molecular orbitals then correspond to the image that the lone pairs on the ligands have donated to the metal ion, the extent of the donation being represented by the α value.

THE STABILITY OF COMPLEX COMPOUNDS

It is desirable to have an estimate of the strengths of the bonds between metal ion and ligand. Convenient means of determining these bond energies include the direct calorimetric measurement of their heats of formation, or perhaps less directly, by measuring the Gibbs free energy of formation (ΔG°) which is usually calculated from the formation constants (or more often called stability constants) K_i by $\Delta G_i^0 = -RT\ln K_i$. K_i is defined for the following equilibrium:

$$ML_i : K_i = [ML_{i-1}/[ML_{i-1}][L] \quad (IV.4)$$

These two experimental monitors of the strength of complex formation have revealed several important principles.

(a) Complexing occurs in a stepwise fashion, one ligand at a time becoming bonded to the metal ion (often by replacing one molecule of solvent water at a time).

(b) Multidentate ligands form more stable complexes than the corresponding monodentate ligands. This, for example, explains the tremendous stability of porphyrin, enzyme and EDTA complexes.

(c) In general, the greater the positive charge density of a metal ion, the more stable is the complex.

(d) For a given set of ligands and metal ions of the first transition series, ligand field effects dictate that the complex stabilities obey the order:

$$Mn(II) < Fe(II) < Co(II) < Ni(II) < Cu(II) > Zn(II)$$

(e) Metal ions can be grouped into two classes; for one class of metal ions called 'hard' acids, the stability of their complexes decreases in the order (as expressed through ligand donor atoms): $F^- > Cl^- > Br^- > I^-$, $O > S$, or $N > P$. On the other hand the stability of metal ions of 'soft' acid character decreases in the opposite order: $I^- > Br^- > Cl^- > F^-$, $S > O$, $P > N$. This represents only general tendencies. There are metal ions on the border line.

LABILITY OF METAL COMPLEXES

If the rate of ligand exchange or substitution of a metal complex is high, the complex is said to be labile to ligand substitution. If the rate is low, the complex is called inert. The exchange of water is very slow in $[Cr^{III}(H_2O)_6]$, whereas it is very fast in $[Cr^{II}(H_2O)_6]$, $[Mn^{II}(H_2O)_6]$, $[Cu^{II}(H_2O)_{4,6}]$, and others. That is, $[Cr^{III}(H_2O)_6]$ is inert, and the others listed are labile. Very generally speaking, Cr(III) complexes and low-spin Co(III) complexes are substitution-inert and the other transition metal and post-transition metal complexes are labile. Inertness should not be confused with stability; the former is a kinetic concept, while the latter is a thermodynamic concept. In certain cases there may be a correlation between the stability and the inertness, but a stable complex can be very labile.

SPECIAL TYPES OF COORDINATION COMPOUNDS

(a) Some metal complexes have no overall charge and can be readily soluble in organic solvents such as the lipid-protein membranes of cells. This is often a convenient means of transporting a metal ion into a cell. When the ligand is large and lipophilic like valinomycin (*see* Experiment 13), the resulting complex may be lipid-soluble despite its ion-pair character.

(b) Hydrolysis (the production of hydroxy complexes) occurs very readily in aqueous solution. This is often through a complexed water molecule losing a proton to give a complexed hydroxide ion. The process can be suppressed in the experiments that follow by maintaining an excess of mineral acid protons in the solution.

(c) Metal ions of oxidation states that would apparently be unstable in aqueous solution can be stabilised by complexing; e.g. $[Cu(H_2O)_4]I_2$ is unstable and soon yields $CuI + I_2$. $[Cu(en)_2]I_2$ (en = ethylenediamine) is quite stable and does not reduce to the cuprous state. Similarly, Fe(II) easily oxidises to Fe(III) but, when complexed, it can be stabilised. This is related to the redox potential of the metal complexes.

(d) The directive nature of coordination bonds gives rise to a whole host of geometric isomers of *cis* and *trans* variety and when ligands are bidentate the range of possibilities is enlarged to encompass optical isomers. Optically active ligands, such as L amino acids, and several different ligands being mixed on the same metal ion gives even more possibilities. It is challenging to find that only selected types of these complexes are active physiologically (e.g. the *cis* isomer of $Pt(NH_3)_2Cl_2$ being an active carcinostatic drug, the *trans* isomer being inactive). The evolution of nature to be dependent upon certain optically active or geometrical isomers is a subject in its own right.

The rapidly developing field of bio-inorganic chemistry is centred upon the existence of coordination compounds in living systems. When one transposes these inorganic principles into biological terms one is met with many complicating factors, not all of which have, as yet, been overcome or understood. The bio-inorganic complexes invariably contain a range of different ligands attached to the same metal ion (compare this to the six equal ligands such as cyanides complexed to Fe(II) in our earlier examples). Thus asymmetric complexes and bonds being in a stretched, or 'entatic', state are far more common. Concentrations are lower in biological systems. Solutions may be non-aqueous such as in the active sites of metalloenzymes or in the membranes of cells. Frequently the energy liberated upon complexing at one metal ion is used to effect a metal-dependent reaction at another site. Nature is far more efficient than laboratory research workers!

For the non-inorganic chemistry student, a clear, readable description of these inorganic principles is given in G. I. Brown (1967), *A New Guide to Modern Valency Theory,* Longmans, London. Inorganic chemists may prefer to read S. F. A. Kettle (1969), *Coordination Compounds,* Nelson, London.

Chapter V

Instrumental Measurements

When one makes physical measurements on a bio-inorganic system a very necessary precursor is the defining of the objectives of the experiment. In some instances we are attempting to answer the question whether complexing is occurring between a metal ion and a given ligand in the biological environment. At other times we know that complexing occurs but need to establish the type of complex – spin paired or spin free? how many ligands are attached to the metal ion? what is the stereochemistry of the complexes formed – *cis* or *trans*, etc.? Finally, one frequently needs to know the exact concentrations of the complexes present. Thus, the selection of the correct instrumental approach depends very much on the terms of reference of the question being tackled. Naturally, not all of the methods available *in vitro* for inorganic chemistry researches can be glibly transposed to biology and applied *in vivo*.

With the possible exception of those researches designed to test whether an organism is dependent upon an element in its diet or not, whole organism studies of bio-inorganic chemistry are not feasible. Instead, we must examine single compounds extracted from these whole organisms. For complex single species such as metalloenzymes these investigations may merit model studies. For simpler species, such as amino-acid complexes, computers can sometimes be used to combine the results of several single species into models of sets of competing equilibria *in vivo*. However, whichever the approach, one has to accept the challenge of separating out the species from the whole organism, of concentrating it and of purifying it, each of these steps running the risk of altering the stereoisomeric conformation away from that which existed in the biological environment.

Table V.1 lists some popular instrumental techniques used to detect inorganic complexes in solution. Many of these are now being used to unravel the intricacies of bio-inorganic solutions so that step by step we are able to narrow down the role played by chemical intuition in 'guestimating' which complexes are present *in vivo*.

Table V.I

Popular Instrumental Techniques for the Detection of Complexes in Solution

(i) *Spectroscopic*

(a) *Visible and ultraviolet spectra* mainly used for detecting transition-metal complexes (ions with 1–9 *d*-electrons only). Colours due to '*d–d*' electron transitions are usually sensitive to changes in the ligands:

$$\text{e.g.}\quad [Cu_{aq}]^{2+} \longrightarrow [CuCl_4]^{2-}$$

blue → yellow

(b) *Infrared spectra* (*not* in aqueous solution)
e.g. frequency of C=O stretching will alter if the 'O' is coordinated to a metal ion as is the case in, for example, acetylacetone complexes (chelates)

(ii) *Conductance measurements* can be used to show a decrease in the number of ions, e.g. $Fe^{3+}_{aq} + 6CN^- \rightleftharpoons [Fe(CN)_6]^{3-}$

(iii) *Transference studies* in the presence of an electric potential, e.g. Fe^{3+} moves towards cathode whereas $[FeCl_4]^-$ moves towards anode; uncharged complexes, e.g. $[Pt(NH_3)_2Cl_2]$ do not move

(iv) *Ion-exchange extraction.* Charged species will be extracted by the appropriate ion exchanger (anion or cation exchanger). Such studies are not always conclusive as complexes may exist in the resin phase and not in the bulk solution, or alternatively a cation-exchange resin might itself complex the metal ion

(v) *Solubility and distribution studies.* Complexes or uncharged ligands are often extractable into a non-aqueous layer, e.g. $H[FeCl_4]$ is extracted into ether from 6 mol dm^{-3} HCl; distribution of NH_3 between water and chloroform is altered if copper(II) is present in the aqueous layer

(vi) *Emf methods.* The emf of an electrolytic cell of the type M | solution of M^{n+} ‖ standard (e.g. calomel) 'electrode' depends upon the concentration of 'free' M^{n+}_{aq} which is required by complex formation (i.e. M + L → ML). Similar cells respond to some anion concentrations, e.g. Cl^- or Br^-, and hence follow free ligand concentration. Nowadays, many 'ion selective electrodes' that respond to specific cations such as Na^+, Ca^{2+}, Cu^{2+}, etc. are available and are widely used

(vii) *pH method* is a special case of (vi) using an H^+ sensitive electrode (e.g. glass electrode). This depends on the fact that many ligands are also bases and hence the pH will fall on complex formation, e.g. $LH^+ + M^{2+} \rightleftharpoons ML^{2+} + H^+$

(viii) *Freezing-point depression and boiling-point elevation.* Measures apparent molecular weight of a complex but is not very useful for aqueous solutions

Every possible analytical technique ought to be exploited to the full and adapted to the special difficulties associated with biological systems. Clearly, not all difficulties and techniques can be exemplified in a dozen or so experiments. However, we have touched upon the three main subsections of this work – the analysis of *total* elemental concentrations (by atomic absorption techniques), the analysis of *free* metal ion concentrations, i.e. that remaining uncomplexed (by pH and ion-selective-electrode potentiometry), and the determination of bonding and structural details of the complexes present (by spectral analyses).

REFERENCES FOR FURTHER READING

A. I. Vogel (1966), *A Textbook of Quantitative Inorganic Analysis,* Longmans, London, 3rd edn.

W. B. Guenther (1968). *Quantitative Chemistry,* Addison Wesley

G. Pass and H. Sutcliffe (1968). *Practical Inorganic Chemistry,* Chapman and Hall, London

G. Marr and B. W. Rockett (1972). *Practical Inorganic Chemistry,* Van Nostrand, London

H. F. Walton (1964). *Principles and Methods of Chemical Analysis,* Prentice Hall, New Jersey

I. M. Kolthoff and E. B. Sandell (1952). *Textbook of Quantitative Inorganic Analysis,* Macmillan, London, 3rd edn.

H. A. Flaschka, A. J. Barnard and P. E. Sturrock (1969). *Quantitative Analytical Chemistry,* Vols. I and II, Barnes and Noble, New York

J. M. D'Auria, A. B. Gilchrist and J. J. Johnstone (1973). *Chemistry and the Environment – A Laboratory Experience,* Saunders

Chapter VI

Laboratory Experiments

Experiment 10
The Preparation, Characterisation and Screening of Pharmaceutically Active Metal Complexes

INTRODUCTION

The last decade has seen an upsurge of interest in trace-element medicine. On the one hand this may involve adjusting the concentration of an essential or beneficial element to a figure that produces optimum health. On the other hand, this might involve administering a compound of an element not normally found in the body to stimulate the body to rid itself of an invading organism such as viruses, bacteria or cancer cells. Compounds of Group VIII in the transition series are particularly effective in this respect. The most widely known metal complex carcinostat is *cis*-dichlorodiammineplatinum(II) whereas substituting water molecules in place of the two chlorides gives a viable antiviral agent. Finally, some negatively charged compounds such as the hexachloroplatinate(IV) anion have marked bactericidal activity. Clearly, we cannot cover all aspects of the preparation, characterisation and screening of these drugs but we can illustrate some aspects.

PROCEDURE

Prepare *cis*-$Pt(NH_3)_2Cl_2$ and purchase commercial samples of *cis*-$Pt(NH_3)_2$-$(NO_3)_2$ and K_2PtCl_6.

Cis-$Pt(NH_3)_2Cl_2$

Platinum nitrate is dissolved in excess aqueous ammonia solution and then this solution is acidified with perchloric acid and allowed to stand for 24 h. A large excess of sodium chloride is added as a concentrated aqueous solution and the product precipitates after a short time. This is filtered, washed with distilled water and dried *in vacuo*. Submit a sample for elemental analyses, record its infrared spectra and measure its paramagnetic moment.

This latter investigation is to establish whether the Pt^{2+} is

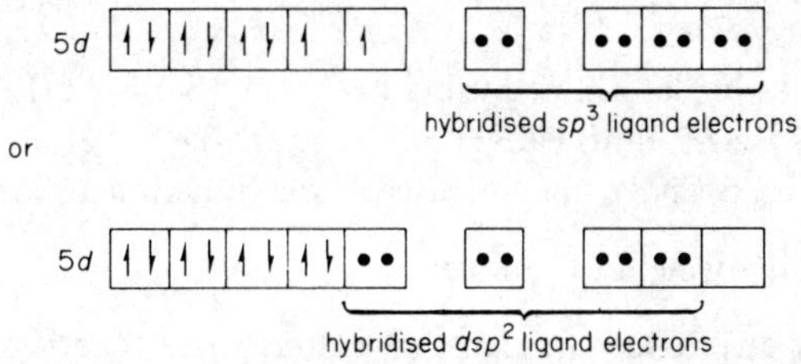

Paramagnetic measurements

These determinations can be made by modifying a conventional single-pan balance to weigh a small magnet with and without the sample being studied between its poles. (This is the converse of the more expensive Gouy method which weighs a sample with and without a magnetic field – *see* Experiment 15). The decrease in weight of the magnet is proportional to the paramagnetic moment of the compound in the tube (*see Figure VI.10.1*).

Measure and tabulate the molar susceptibilities (χ_M), the mass susceptibilities (χ), the magnetic moments (μ), the number of unpaired electrons that you would expect (n) and the theoretical spin magnetic moments (μ') for the three compounds listed at the beginning of this experiment. What do you conclude concerning the arrangement of electrons in your complexes?

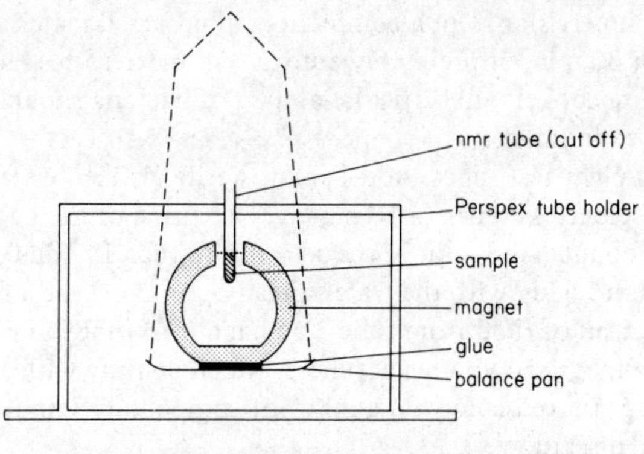

Figure VI.10.1 Conversions to a single-pan balance to permit magnetic moment determinations

The basic equations required are:

$$\chi_m = \frac{C'M\Delta_W}{W} \quad \text{(i)}$$

$$\chi = \chi_{m/M} \quad \text{(ii)}$$

$$\mu = 2.83\,(\chi'_m \cdot T)^{1/2}\ \text{B.M.}\ \ (\chi'_m \text{ in c.g.s. units}) \quad \text{(iii)}$$

$$\mu' = (n(n+2))^{1/2} \quad \text{(iv)}$$

where C' = constant to be determined using $Fe(NH_4)_2(SO_4)_2 . 6H_2O$ for which $\chi = 3.26 \times 10^{-5}$ c.g.s. units (409.7×10^{-12} m^3 kg^{-1}) at 20°C or, alternatively, $HgCo(CNS)_4$ may be used – *see* Experiment 15;

M = molecular weight of sample;

Δ_W = weight difference of magnet *or* sample when together and separate;

W = weight of sample (choose minimum volume to give maximum Δ_W);

T = absolute temperature;

and χ'_M = χ_M + a diamagnetic susceptibility correction – for the purposes of this experiment $\chi'_M = \chi_M$

Antibacterial activity

Materials required: (a) freeze-dried culture of *Esherichia coli* ATCC 11229; (b) a McCartney bottle containing 25 ml of nutrient broth; (c) sterilised, capped, disposable test tubes (24 of 100 mm x 15 mm); (d) sterile distilled water (50 ml); (e) sterile disposable graduated pipettes (30 of 1 ml).

Safety precautions must be observed. No bacterium can be guaranteed to be harmless under all circumstances and so the experiment described should only be carried out under supervision. Upon completion, all infected material should be boiled in water for 30 min, or preferably autoclaved under 15 p.s.i. of steam (a domestic pressure cooker will suffice). Samples of bacteria should never be used for a second experiment.

Set up a row of eight test tubes. Add 1 ml of sterile distilled water to tubes 2-8 using a sterile pipette. Prepare a 10 mg ml^{-1} solution of the hexachloroplatinate(IV) compound. Add 1 ml of the solution to tube 1. Add 0.5 ml to tube 2, and mix thoroughly with the water already present. Take a fresh pipette, remove 0.5 ml of fluid from tube 2 and add it to tube 3, mix thoroughly. Proceed in this manner until you reach tube 8, which ends up with 1.5 ml of solution containing approximately 3 μg ml^{-1} of compound. Remove 0.5 ml from this tube and discard.

Take 0.5 ml of your culture of *E. coli* and add it to the 25 ml of nutrient broth. Mix thoroughly, then add 1 ml of this dilute bacterial suspension to each tube, starting with those containing dilute solutions of compound, and gradually working toward those with concentrated solutions. Provided you do not allow the pipette to touch the liquid in the tubes you can use the same one for all additions, but you must use a fresh one if the pipette accidentally touches the contents of a tube. Mix thoroughly and note which tubes are cloudy.

Place in a 37°C incubator overnight or leave for 3 days in the laboratory. Some of the tubes which were clear will have gone cloudy, indicating growth of bacteria. Prepare a table of results, marking '+' where a tube has gone cloudy and '–' where it has remained clear.

SUGGESTIONS

Either write an essay reviewing the inorganic principles involved in the design of platinum anticancer drugs *or* write a short precis of current theories concerning the mechanism through which *cis*-$Pt(NH_3)_2Cl_2$ exhibits its anticancer activity.

REFERENCES

B. Rosenberg (1971). 'Some Biological Effects of Platinum Compounds'. *Plat. Metals Rev.*, 42

D. R. Williams (1972). 'Metals, Ligands and Cancer'. *Chem. Rev.*, **72**, 203

D. R. Williams (1972). 'Anticancer Drug Designs involving Complexes of Amino-acids and Metal Ions'. *Inorg. Chim. Acta. Rev.*, **6**, 123

M. J. Cleare (1974). 'Transition Metal Complexes in Cancer Chemotherapy'. *Coord. Chem. Rev.*, **12**, 349

M. Axford, A. Davies, D. Forbes and D. Wilson (1974). *Antibacterials*, I.C.I. Publication, Kynoch Press

P. W. Selwood (1956). *Magnetochemistry*, Interscience

Experiment 11
Nickel(II)—Glycine Interactions Studied by pH Potentiometry

INTRODUCTION

The scientific literature contains an increasing number of reports of computer models of biological fluids. These models give caracatures of the concentration and pH-dependence of complexes present *in vivo* and reveal, for example, which essential complexes are going to suffer from such clinical situations as a deficiency of an essential metal or the presence of a polluting metal (*see* Experiment 25). There are three basic requirements for such models – powerful computer programs, a knowledge of the total ingredients in a body fluid (as from Experiments 1-4), and the values of the formation constants, β, for all the important complexing reactions occurring in the fluid under examination. This experiment shows how pH potentiometry may be used to establish these β values. The nickel(II)-glycinate system is used as illustration.

THEORY

Glycine exists as a zwitter ion, $^{+}NH_3CH_2COO^{-}$, HL, and associates in an aqueous solution as follows:

$$L^- + H^+ \rightleftharpoons HL \qquad (11.1)$$

$$K_H = [HL]/[H^+][L^-] \qquad (11.2)$$

K_H, or $\log K_H$, can be evaluated by titrating a glycine solution with a standardised NaOH solution and monitoring the extent of protonation by using a glass electrode to measure some function of $[H^+]$.

Similarly, the amount of ligand complexed to a metal ion (equations 11.3, 11.4 and 11.5) can also be monitored using a glass electrode because $[L^-]$ in equations 11.3 to 11.5 is related to $[H^+]$ through equation 11.1 and K_H:

$$Ni^{2+} + L^- \overset{K_1}{\rightleftharpoons} NiL^+ \qquad (11.3)$$

$$NiL^+ + L^- \overset{K_2}{\rightleftharpoons} NiL_2 \qquad (11.4)$$

$$NiL_2 + L^- \overset{K_3}{\rightleftharpoons} NiL_3^- \qquad (11.5)$$

Electrodes respond to the activities (a) of species in solution (e.g. a glass/reference electrode pair monitors a_{H^+}), these being products of activity coefficients (γ) and concentrations ([ ]), $a = \gamma.$ [ ]. Constants such as K_H or K_1 expressed in terms of activities are called 'thermodynamic constants'. These 'constants' are only applicable at very low ionic strength and when this changes (e.g. $Ni^{2+} + L^- \rightarrow NiL^+$) the activity coefficients change. A convenient means of avoiding this problem is to use 'concentration constants' defined solely in terms of [ ] and measured in the presence of a background electrolyte sufficiently concentrated to hold γ values constant even if charges (and therefore ionic strengths) are changing during the complexing reactions. Bio-inorganic chemists often use a background electrolyte equivalent to 150 mmol dm^{-3} NaCl since this approximates to conditions in blood plasma. Electrode pairs are then calibrated not in terms of activities (using buffers) but in terms of concentrations using acid solutions for which the $[H^+]$ is known. This approximates to conditions in blood plasma and produces the E_0 for the electrode

$$E = E_0 + 59.16 \log [H^+] \text{ mV at } 25.0°C$$

E_0 is constant and can be used in all subsequent experiments for which E is measured and $[H^+]$ is being determined.

Returning to the problem of evaluating K_H for glycine, a solution of glycine (150 mmol dm^{-3}) is titrated with (100 mmol dm^{-3}) NaOH solution. With each addition of NaOH, the emf of the electrodes is measured and using E_0, E is converted into $[H^+]$. In such solutions, several conditions must hold:

(i) $[H^+] + [Na^+] = [OH^-] + [L^-]$ (for electric neutrality) (11.6)

(ii) the total concentration of glycine, $L_{tot} = [HL] + [L^-]$ (11.7)

Combining equation 11.2 with equations 11.6 and 11.7 we get the following equation

$$\log K_H = -\log [H^+] + \log \left\{ \frac{L_{tot} - ([Na^+] + [H^+] - [OH^-])}{[Na^+] + [H^+] - [OH^-]} \right\} \tag{11.8}$$

This is the exact formulation for a pK_H titration; however, in near-neutral solutions, $[H^+]$ and $[OH^-]$ are small compared to $[Na^+]$ and may be neglected and so we obtain the simplified equation:

$$\log K_H = -\log [H^+] + \log \left\{ \frac{L_{tot} - [Na^+]}{[Na^+]} \right\} \tag{11.9}$$

All the quantities on the right-hand side can be calculated from E and the concentrations of reactants in titrant and titrate. (Note that $[Na^+]$ is calculated from the amount of NaOH added and the $[Na^+]$ arising from the background electrolyte is ignored. Why is this justified? *See* equation 11.6.)

Formation curves for the nickel–glycinate system

In any solution of glycine and nickel ions the following equilibria occur:—

$$NH_2.CH_2COO^- + H^+ \rightleftharpoons NH_3^+.CH_2.COO^- \tag{11.10}$$

$$\begin{aligned} NH_2.CH_2COO^- + Ni^{2+} &\rightleftharpoons NH_2.CH_2.COO.Ni^+ \\ 2(NH_2.CH_2COO^-) + Ni^{2+} &\rightleftharpoons (NH_2.CH_2.COO)_2Ni \\ 3(NH_2.CH_2.COO^-) + Ni^{2+} &\rightleftharpoons (NH_2.CH_2.COO)_3.Ni^- \end{aligned} \tag{11.11}$$

If we measure the $[H^+]$ of reaction 11.10 we are, in fact, obtaining a quantitative measure of how much of reactions 11.11 has occurred. Thus, gradually changing the total acid content of the system can take the solution through the spectrum of complexes from $NH_3^+.CH_2.COO^-$ to

$(NH_2.CH_2.COO^-)_3Ni^{2+}$, and furthermore, potentiometrically measuring the free $[H^+]$ will tell us the exact extent of this change.

The results of the titration are expressed as a formation curve, i.e. a plot of $\overline{Z}$ versus $-\log[NH_2.CH_2.COO^-]$; $\overline{Z}$ is the average number of glycinates bound per Ni^{2+} and $-\log[L^-]$, or pA for short, is the 'push' we are putting into reactions 11.11. Note that we speak of $[NH_2.CH_2.COO^-]$ and not $[NH_2.CH_2.COOH]$. This is because the Ni^{2+} and the H^+ are competing with each other for the $[NH_2.CH_2.COO^-]$.

The formation curve is calculated by using a computer program, the mathematics of which are as follows.

If $L^- = NH_2.CH_2.COO^-$, total ionisable protons present $T_H = [H^+] + [HL]$
and $\beta_1 = K_H = [HL]/([H^+][L^-])$
Thus $T_H = [H^+] + [H^+][L^-].\beta_1$
and $[L^-] = (T_H - [H^+])/(\beta_1[H^+])$, hence $-\log[L^-] = pA$
Similarly, total L, $T_L = [L^-] + [HL] + (\text{L that is complexed})$

$$\text{Now } \overline{Z} = \text{average number of L per } Ni^{2+} = \frac{(\text{L that is complexed})}{T_{Ni^{2+}}}$$

$$= (T_L - [L^-](1 + \beta_1[H^+]))/T_{Ni^{2+}} = \overline{Z}$$

Plot $\overline{Z}$ against pA.

In order to measure as wide a range of T_H as possible, we adopt the titration procedure and so can measure about 60 points on the formation curve in a day. A computer program is available for turning the measured emfs into $\overline{Z}$/pA values (*see* pages 54 and 55). (Consult a demonstrator about how to use the computer and its programs.)

The terms used in the computer program are as follows:

M	= Number of sets of titrations to be computed.
VINIT	= Initial volume on vessel (in ml).
HSI	= Mineral acid concentration in solution S, initially. Note that we titrate solution T into solution S. Mineral acid = potentially ionisable protons, e.g. from HL, or HCl used to prepare nickel solution.
HTI	= Mineral acid concentration in solution T, initially.
ASI	= Ligand concentration in solution S, initially.
AT	= Ligand concentration in solution T, initially.
E0	= E_0 for the electrode pair.
BS	= Metal concentration in S.
BT	= Metal concentration in T.
BONE	= β_1.
WK	= K_W (ionic product for H_2O).

VADD(I) = Volume added (in ml) for i^{th} point.
E(I) = Emf for any point i.
AA(I) = T_A.
FLIG(I) = $[A]_i$.

Other terms ought to be self-evident from the mathematics just given.

Formation constants from the formation curve

Volumes have been published concerning the best methods of describing formation curves with constants but as far as the present experiment is concerned the simplest, and fastest, Bjerrum $Z_{½}$ method is used.

This method is best described by analogy with the p*K*s of acids and bases (elementary undergraduate chemistry courses), e.g. CH_3COOH.

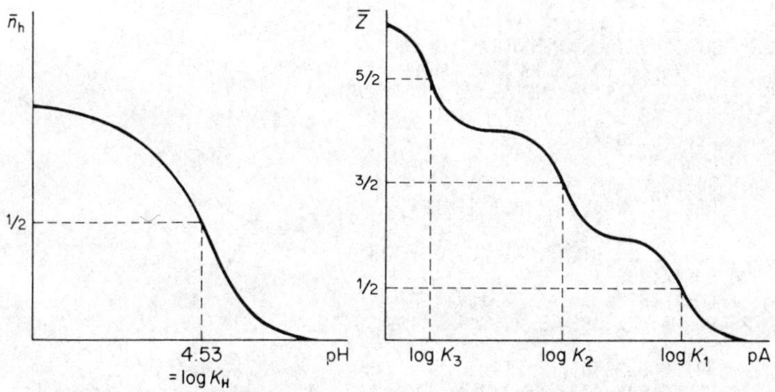

$\bar{n}_h$ = average number of H^+ per CH_3COO^-. At pH 9, all acetate is CH_3COO^-. When $[CH_3COOH]$ = $[CH_3COO^-]$, i.e. $\bar{n}_h$ = ½, log K_H = pH as per Henderson's equation.

$\bar{Z}$ = average number of L^- per metal ion. Bjerrum's method shows mathematically (*see* References) that at $\bar{Z}$ = ½, log K_1 = pA and at $\bar{Z}$ = 3/2, log K_2 = pA etc.

PROCEDURE

The electrode pair is calibrated by measuring the emf in the solutions of known $[H^+]$, ionic background = 150 mmol dm^{-3} NaCl, provided and applying the Nernst relationship.

Log K_H determination

Prepare the following solutions: 25 ml of 100 mmol dm^{-3} glycine, 500 ml of 300 mmol dm^{-3} NaCl and 100 ml of 100 mmol dm^{-3} NaOH (by diluting ampoule NaOH with 50 ml of 300 mmol dm^{-3} NaCl and water).

Add the magnetic stirrer bar and introduce the calibrated electrode pair into the titration vessel containing 5 ml of water, 15 ml of 300 mmol dm^{-3} NaCl and 10 ml of 100 mmol dm^{-3} glycine. Fill the burette with 100 mmol dm^{-3}

```
C     AN UNDERGRADUATE PROGRAM FOR STABILITY CONSTANTS FOR THE NICKEL(II)
C    GLYCINATE SYSTEM
C
C
C     M IS NUMBER OF TITRATIONS
C     VINIT IS INITIAL VOLUME IN VESSEL
C     HSI IS ACID CONCENTRATION IN SOLUTION S ORIGINALLY
C     HTI IS ACID CONCENTRATION IN SOLUTION T ORIGINALLY
C     ASI IS LIGAND CONCENTRATION IN SOLUTION S ORIGINALLY
C     AT IS LIGAND CONCENTRATION IN SOLUTION T ORIGINALLY
C     BS IS METAL CONCENTRATION IN S
C     BT IS METAL CONCENTRATION IN T
C     WK IS IONIC PRODUCT OF WATER
C     VADD(I) IS VOLUME ADDED FOR POINT (I)
C     E(I) IS EMF FOR ANY POINT (I)
      READ(5,10)M
   10 FORMAT(I2)
      DO 57 J=1,M
      READ(5,14)BS,ASI,HSI,VINIT
   14 FORMAT(3E11.4,F7.3)
      READ(5,20)BT,AT,HTI,EO,BONE,WK
   20 FORMAT(3E11.4,F7.2,2E11.4)
      WRITE(6,25)BS,ASI,HSI,VINIT
   25 FORMAT(4X,'BS=',E11.4,1X,'ASI=',E11.4,1X,'HSI=',E11.4,1X,'VINIT='.
     2F7.3//)
      WRITE(6,28)BT,AT,HTI,EO,BONE,WK
   28 FORMAT(4X,'BT=',E11.4,1X,'AT=',E11.4,1X,'HTI=',E11.4,1X,'EO=',F7.2
     3,1X,'BONE=',E11.4,1X,'WK=',E11.4//)
      WRITE(6,32)
   32 FORMAT(2X,'VADD',6X,'E',6X,'PH',8X,'ZBAR',5X,'PA',10X,'FLIG'/)
      DIMENSION V(100),TH(100),AA(100),PH(100),H(100),ZONA(100),ZBAR(100
     4),PA(100),FLIG(100),VADD(100),B(100),E(100)
      I=1
   39 READ(5,35)VADD(I),E(I),INDEX
C     INDEX IS ZERO FOR EVERY POINT IN A TITRATION EXCEPT THE LAST ONE
   35 FORMAT(2F10.3,4X,I5)
      V(I)=VINIT+VADD(I)
      TH(I)=HSI*VINIT)/V(I)+(HTI*VADD(I))/V(I)
      PH(I)=(EO-E(I))/59.16
      AA(I)=(ASI*VINIT/V(I))+(AT*VADD(I)/V(I))
      B(I)=(BS*VINIT/V(I))+(BT*VADD(I)/V(I))
      H(I)=EXP(-2.3026*PH(I))
      FLIG(I)=(TH(I)-H(I)+WK/H(I)/(BONE*H(I))
      ZBAR(I)=(AA(I)-FLIG(I)*(1.0+BONE*H(I)))/B(I)
      IF(FLIG(I))40,40,42
   40 PA(I)=0.0
      GO TO 45
   42 PA(I)=-((ALOG(FLIG(I)))/2.3026)
   45 ZONA(I)=ZBAR(I)/FLIG(I)
      WRITE(6,48)VADD(I),E(I),PH(I),ZBAR(I),PA(I),FLIG(I)
   48 FORMAT('0',F7.3,2X,F10.3,2X,F7.4,3X,F7.4,3X,F8.4,3X,E11.4/)
      I=I+1
      IF(INDEX.EQ.0)GO TO 39
   57 CONTINUE
      STOP
      END
```

NaOH and add about 15 aliquots to the titration vessel to titrate 0–100% of the glycine. For each point record the volume added and the emf reading. Calculate log K_H using equation 11.9 for points in 20–80% of the titration range.

Stability constant determination

You are provided with $NiCl_2$ solution containing acid to suppress hydrolysis. Prepare a solution of sodium glycinate, 250 ml of 100 mmol dm^{-3} (weighed glycine + ampoule NaOH from protonation experiment + NaCl solution to 150 mmol dm^{-3}).

Place 20 ml of 300 mmol dm^{-3} NaCl and 20 ml of 10 mmol dm^{-3} $NiCl_2$ in the titration vessel, stir and add the sodium glycinate in 0.20 ml aliquots, recording volume and emf for each addition.

Feed these data into the computer program and estimate log K_1, log K_2 and log K_3 from the plotted output. Compare these results with those listed in compilations of stability constants in the reference section of your library.

Data for Nickel Glycinate Stability Constants Program

Card No. 1. Number of titrations, Format I2

```
|?|
```

2. Metal, ligand and ionisable protons (mol dm^{-3}) in solution in vessel; initial volume (ml) in vessel: Format 3E11.4, F7.3

```
|±|?|·|?|?|?|?|E|±|?|?|±|?|·|?|?|?|?|E|±|?|?|±|?|·|?|?|?|?|E|±|?|?|±|?|?|·|?|?|?| | |
```

3. Metal, ligand, ionisable protons (mol dm^{-3}) in burette, electrode E^0, β_1 and K_w for water; Format 3E11.4, F7.2, 2E11.4

```
|±|?|·|?|?|?|?|E|±|?|?|±|?|·|?|?|?|?|E|±|?|?|±|?|·|?|?|?|?|E|±|?|?|±|?|?|?|·|?|?|±|?|·|?|?|?|?|E|±|?|?|±|?|·|?|?|?|?|E|±|?|?|
```

4. Volume added (ml), emf and index (the index is zero for every point in a titration except the last one); Format 2F10.3, 4X, I5

```
| | | | |?|?|·|?|?|?| | | | | |·|?|?|?| | | | |
| | | | |?|?|·|?| |?| | | | | |·|?|?|?| | | | |
                etc.
                etc.
| | | | |?|?|·|?|?|?| | | | | |·|?|?|?| | | |1|   (last card)
```

REFERENCES

D. R. Williams (1971). 'The Uranyl Acetate System Studied by pH Potentiometry'. *J. Chem. Educ.*, **48**, 480. (Experimental Measurements)

S. Glasstone (1955). *Physical Chemistry,* Macmillan, 2nd edn., 1003. (Henderson's Equation)

W. B. Guenther (1968). *Quantitative Chemistry,* Addison Wesley, 222. (Bjerrum's Method)

W. B. Guenther. *J. Chem. Educ.,* **44**, 46 (1967); *Quantitative Chemistry,* Addison Wesley, 206 (1968). (Formation Curves)

D. D. Perrin and R. P. Agarwal (1973). 'Multimetal-Multiligand Equilibria: A Model for Biological Systems'. *Metal Ions in Biological Systems, Vol. 2,* Ed. H. Sigel, Dekker (Computer Models)

F. J. C. Rossotti, H. S. Rossotti and R. J. Whewell (1971). 'The Use of Electronic Computing Techniques in the Calculation of Stability Constants'. *J. Inorg. Nucl. Chem.,* **33**, 2051. (Superior Methods of Converting Formation Curves into βs)

Stability Constants, The Chemical Society, London, Spec. Publs. Nos. 17 and 25 (1964 and 1971). *Critical Stability Constants,* Eds. A. E. Martell and R. M. Smith, Plenum, New York (1974). (Compilations of Stability Constants)

R. J. Angelici (1969). *Synthesis and Technique in Inorganic Chemistry,* Saunders, Experiment number 13. (More Exacting Methods of Obtaining $\log K_1$, $\log K_2$, etc. from Formation Curves)

Experiment 12
Determination of Formation Constants of Calcium – ATP Complexes

INTRODUCTION

Adenosinetriphosphate (ATP) is one of the most important energy carriers in biological systems. The hydrolysis of ATP to form adenosinediphosphate (ADP) and orthophosphate provides energy for numerous biological activities such as muscle contraction, transmembrane ion transport and synthesis of macromolecules.

Owing to its many functional groups, ATP^{4-} can complex with metal ions in various ways. ATP^{4-} exists mostly chelated to Mg(II) in cell fluid. Mg(II) is required for the activity of various enzymes such as ATPases which catalyse the hydrolysis of ATP and kinases and the transfer of a phosphate group from ATP to substrates. Na(I) and K(I) are transported across cell membranes by a mechanism called a 'sodium pump', the essential portion of which is performed by a special ATPase activated by Na(I), K(I) and Mg(II) ions.

Likewise, Ca(II) ion is transmembraneously carried by a 'calcium pump', which is essentially a Ca(II)/Mg(II)-activated ATPase embedded in membrane. Ca(II) is known to trigger muscle contraction and may involve itself in an interaction with ATP.

In this experiment, the interaction between ATP and Ca(II) ion is studied quantitatively. The activity and hence the concentration of free Ca(II) ion is determined by a Ca(II)-specific electrode.

THEORY

The equilibrium of a system containing Na(I), Ca(II) and ATP^{4-} is described by

$$ATP^{4-} + Na(I) \overset{K_{Na}}{\rightleftharpoons} NaATP^{3-} \quad (12.1)$$

$$ATP^{4-} + Ca(II) \overset{K_1}{\rightleftharpoons} CaATP^{2-} \quad (12.2)$$

$$Ca(II)^{-} + CaATP^{2-} \overset{K_2}{\rightleftharpoons} Ca_2ATP^0 \quad (12.3)$$

The equilibrium constants are

$$K_{Na} = a_{NA}/a_A.a_N = [Na]\gamma_3/[A]\gamma_4[N]\gamma_1 \; (= 229 \text{ dm}^3 \text{ mol}^{-1}) \quad (12.4)$$

$$K_1 = a_{CA}/a_A a_C = [CA]\gamma_2/[A]\gamma_4[C]\gamma_2 \quad (12.5)$$

$$K_2 = a_{C_2A}/a_C a_{CA} = [C_2A]/[C]\gamma_2[CA]\gamma_2 \quad (12.6)$$

where a_i is the activity of species i, γ_z is the activity coefficient of a species whose charge is $\pm z$, and N = Na(I), NA = $NaATP^{3-}$, C = Ca(II), CA = $CaATP^{2-}$, $C_2A = Ca_2ATP^0$. According to Debye-Hückel theory, γ_z is related to the ionic strength μ of the solution in the following way:

$$\log \gamma_z = -0.5091 z^2 \mu^{1/2}/(1 + \mu^{1/2}) \quad (12.7)$$

If this approximation is used, equations 12.4–12.6 may be re-written as

$$K_{Na} = [NA]/[N][A]\gamma^2 \quad (12.8)$$

$$K_1 = [CA]/[C][A]\gamma^4 = [CA]/a_C[A]\gamma^3 \quad (12.9)$$

$$K_2 = [C_2A]/[C][CA]\gamma^2 = [C_2A]/a_C[CA]\gamma \quad (12.10)$$

where $\gamma = \gamma_2$. That is, all K_i values can be expressed in terms of γ and others.

When the ratio of the total calcium concentration, T_C, to the total ligand (ATP) concentration, T_A, is $\leqslant 1.0$, it can be safely assumed that $CaATP^{2-}$ is the only Ca–ATP complex present in the system. Under this restriction, the formation constant, K_1, will be determined. The material balances are represented by:

$$T_C = [CaATP^{2-}] + [Ca(II)] \quad (12.11)$$

$$T_A = [ATP^{4-}] + [CaATP^{2-}] + [NaATP^{3-}] \quad (12.12)$$

In the preparation of the sample solution, Na_2H_2ATP and NaOH are used; therefore

$$T_N \text{ (total concentration of Na(I))} = 2T_A + T_{NaOH}$$
$$= [Na(I)] + [NaATP^{3-}] \tag{12.13}$$

a_C is measured directly by a calcium-specific electrode in this experiment. Then $[CaATP^{2-}]$ should be calculated by

$$[CA] = T_C - [C] = T_C - a_C/\gamma \tag{12.14}$$

$[ATP^{4-}]$ should be obtained by use of equation 12.11; i.e.

$$[A] = T_A - [CA] - [NA] \tag{12.15}$$

However, it is not possible to calculate these out until γ is known. Hence, the ionic strength, μ, must be known first. To calculate μ we must know [CA], [A], etc. Thus, it seems impossible to do the whole calculation.

To resolve this dilemma the method of successive approximations is utilised. An initial assumption is made that $\gamma = 1$ and $[NA] = 0$. Using these values, the first approximate values for the other concentrations and hence μ are obtained:

$$[N] = 2T_A + T_{NaOH} - [NA] = 2T_A + T_{NaOH} \tag{12.16}$$

$$[CA] = T_C - a_C/\gamma = T_C - a_C \tag{12.17}$$

From equations 12.8 and 12.12, the following equation is derived:

$$[NA] = (T_A - [CA])/(1 + 1/[N]\gamma^2 K_{Na}) \tag{12.18}$$

Using [N] and [CA] estimated above and equation 12.18 the first approximate value of [NA] is obtained. Now this [NA] value and equation 12.15 are used to evaluate [A]. All these first approximate values are put in the following equation to obtain the first approximate μ value:

$$\mu = (1/2)\{16[A] + 9[NA] + 4[CA] + 4[C] + [NO_3^-] + [OH^-] + [N]\} \tag{12.19}$$

$[OH^-]$ can be evaluated from the pH reading of the solution. $Ca(NO_3)_2$ is used in the experiment; therefore $[NO_3^-] = 2T_C$.

Using the first approximate μ value, γ is now re-evaluated. Then the second approximate values of [CA], [N], [NA] and [A] are calculated. Hence, the second approximate value of μ is established. This procedure is repeated until the differences in μ or $\log \gamma$ between successive steps becomes insignificant. In conclusion, K_1 is evaluated using the final values of $\mu(\gamma)$, [CA], [A], the experimental value of a_C and equation 12.9.

In a solution in which the ratio T_C/T_A is greater than 1, an additional complex Ca_2ATP^0 can be present. In this case, the material balances are represented by:

$$T_C = [Ca(II)] + [CaATP^{2-}] + 2[Ca_2ATP^0] \tag{12.20}$$

$$T_A = [ATP^{4-}] + [CaATP^{2-}] + [Ca_2ATP^0] + [NaATP^{3-}] \qquad (12.21)$$

$[NaATP^{3-}]$ has been found to be negligible under the experimental conditions used. From equations 12.20 and 12.21 the following equations are derived:

$$[CA] = 2T_A - T_C + a_C/\gamma - 2[A] \qquad (12.22)$$

$$[C_2A] = T_C - T_A - a_C/\gamma + [A] \qquad (12.23)$$

An initial assumption is made that $\gamma = 1$ and $[A] = 0$. Now $[CA]$ is estimated from equation 12.22. Equation 12.9 and this $[CA]$ value are utilised to obtain the first approximate $[A]$ value. Then the first approximate value of μ and hence of γ is calculated by use of equation 12.19. This procedure is repeated until a satisfactory agreement is obtained between two successive approximations. Finally, K_2 is evaluated by use of equation 12.23 (for the estimation of $[C_2A]$) and equation 12.10.

METHOD

An Orion 92-20 calcium electrode in conjunction with an Orion 90-02 double junction reference electrode is used in this experiment to measure the activity of the free calcium ion in solutions containing calcium ion and adenosine triphosphate. From this the formation constants of $CaATP^{2-}$ and Ca_2ATP^0 are determined. The electrode responds specifically to free Ca(II) ion in the same way that a conventional glass electrode is specific to H^+ ion. Instead of developing an electrical potential across a glass membrane, the calcium electrode develops the potential across a thin layer of water-immiscible liquid ion exchanger held by an inert porous membrane disc.

Down to 10^{-4} mol dm^{-3} of Ca(II) concentration, the electrode potential can be approximated by the Nernst equation

$$E = E_x + (RT/nF)\ln a_i = E_x + (0.0592/n)\log a_i \text{ (volts at 25°C)} \qquad (12.24)$$

where

E = the electrode potential;

E_x = a constant component of the electrode potential;

n = the charge of the ion involved;

$a_i = \gamma_i \times$ concentration = activity.

Since the electrode behaviour may deviate slightly from Nernstian it is more accurate to relate the potential difference to the activity by means of a calibration curve. A calibration curve can be prepared by making potential measurements on $Ca(NO_3)_2$ solutions of known concentrations ranging from 10^{-5} to 10^{-3} mol dm^{-3}.

The formation constants of $CaATP^{2-}$ and Ca_2ATP^0 can be determined by potentiometric titration of a $Ca(NO_3)_2$ solution to which Na_2H_2ATP is added. In order to determine the formation constant of $CaATP^{2-}$, the T_C/T_A ratios should be kept equal to unity. To determine the formation constant of Ca_2ATP^0 the T_C/T_A ratios will be varied from about 3 to about 1.2.

Potential measurements are made on solutions equilibrated at 25°C and the pH of the sample solution is adjusted to 9.18 by the addition of an NaOH solution.

PROCEDURE

Prepare the following solutions using distilled water:
(i) 250 ml of exactly 1×10^{-2} mol dm^{-3} $Ca(NO_3)_2$;
(ii) 250 ml of exactly 1×10^{-3} mol dm^{-3} $Ca(NO_3)_2$ (by dilution of 1×10^{-2} mol dm^{-3});
(iii) a solution to be shared by all groups consisting of 100 ml of exactly 1×10^{-2} mol dm^{-3} $Na_2H_2ATP.nH_2O$.

50.0 mmol dm^{-3} NaOH solution and 4.00 mmol dm^{-3} $NaNO_3$ solution are provided.

The potentiometer is equipped with three electrodes – a calcium-specific electrode, a glass electrode, and a double junction reference electrode. A switching box connects to the meter inputs for either calcium electrode versus reference electrode for measurements of Ca^{2+} activity, the glass electrode versus reference electrode to permit adjustment of pH, or a direct short circuit to allow adjustment of the meter to 0.0 mV potential difference between the inputs.

Remove the calcium electrode from the holder and clamp it upright to the stand (this electrode must *NOT* be put into the buffer solution). Rinse the remaining electrodes and the cell with water, then pour into the cell a small amount of pH 9.18 buffer. Switch the box to 'calibrate', turn the meter to the mV 'EXP' scale and check and, if necessary, adjust the meter to 0.00. Switch to H^+ measurement and record the mV reading equivalent to pH 9.18 (do not attempt to read pH directly by switching the meter to a pH scale). For all readings the stirring rate should be slow and constant.

Empty the cell with the aspirator, return the calcium electrode to the holder and rinse electrodes and cell with water. Fill one 10-ml burette with 10^{-3} mol dm^{-3} $Ca(NO_3)_2$ and another with 5×10^{-2} mol dm^{-3} NaOH. Pipette 50 ml of 4 mmol dm^{-3} $NaNO_3$ into the cell, and add a 0.50-ml aliquot of 10^{-3} mol dm^{-3} $Ca(NO_3)_2$ solution. Switch to H^+ measurement and carefully adjust the pH with NaOH solution to the previously determined mV reading (± 2 mV). Note that for some solutions *less* than one drop will be required. Switch to the calcium electrode and record the reading when it stabilises (this will be slow for the first reading but faster for all others). Before recording the reading, re-check the pH and also make sure no bubbles are under the calcium electrode. Record the volumes of NaOH and $Ca(NO_3)_2$ added. This procedure is to be repeated for the solution with added titrants as indicated below up to solution 10. Solutions 11–18 are a separate titration. Empty, rinse and clean the cell. Pipette 50 ml of water into the cell and take readings on solutions 11–18 as before.

Solution	Total vol. 10^{-3} mol dm^{-3} Ca^{2+} added (ml)	Total vol. 10^{-2} mol dm^{-3} Ca^{2+} added (ml)	Total vol. 10^{-2} mol dm^{-3} ATP^{4-} added (ml)	[NaOH] for pH adjusting (mol dm^{-3})
1	0.5	–	–	5×10^{-2} in all cases
2	1.0	–	–	
3	2.0	–	–	
4	4.0	–	–	
5	6.0	–	–	
6	10.0	–	–	
7	10.0	1.0	–	
8	10.0	2.0	–	
9	10.0	3.0	–	
10	10.0	4.0	–	
11	10.0	4.0	2.0	
12	10.0	4.0	2.5	
13	10.0	4.0	3.0	
14	10.0	4.0	3.5	
15	10.0	4.0	4.0	
16	10.0	4.0	5.0	
17	10.0	9.0	10.0	
18	10.0	14.0	15.0	

When finished, rinse the cell and electrodes, re-fill the cell with a $Ca(NO_3)_2$ solution (concentration unimportant), and immerse the electrodes.

CALCULATIONS

A calibration curve for the Ca(II) electrode will be made using data from runs containing no ATP (1-10). Determine the ionic strength μ including all ionic species. Find γ (= γ_2) using equation 12.7 and use this to determine a_{Ca}. Plot accurately mV versus log a_{Ca}. Draw the best curve through the points as the electrode response may not be quite Nernstian.

Use the calibration curve to determine a_{Ca} for the remaining points (11–18). Tabulate a_{Ca} for these points along with T_A, T_{NaOH}, T_C and $[OH^-]$. For solutions 16–18, T_C/T_A = 1 and the assumption is made that $CaATP^{2-}$ is by far the most prevalent Ca-ATP complex. Do a sample determination of K_1 for one of these points to the second stage of approximations. Since the solutions are quite dilute, assume initially that μ = 0 and [NA] = 0. (Any other guess would eventually give exactly the same answer but the calculation would take longer.) Using the experimentally determined a_{Ca} calculate [CA] from equation 12.14 and [NA] from equation 12.18. Then approximate [A] (equation 12.12 or 12.15) and re-calculate γ (= γ_2). Repeat the above calculation to determine the second approximations and a rough value of K_1. A calculator or computer can be programmed to use the above tabulated concentrations and activities to give more accurate values of μ, [A] and K_1.

For solutions 11-15 $T_C/T_A > 1$ and thus [Ca_2ATP^0] becomes significant. Assume initially that $\gamma = 1$ and [A] = 0 and calculate K_2 to a second approximation for one of these solutions. The calculator program will give values of [CA], [C_2A] and K_2 for these points. Calculate the average value and standard deviation of K_2.

REFERENCES

E.-I. Ochiai (1977). *Bio-inorganic Chemistry: An Introduction,* Allyn and Bacon, Boston, Chapters 14 and 15. (Biochemistry of Calcium)

H. J. Schatzman (1975). *Current Topics in Membranes and Transport, Vol. 6,* 125. (Calcium Pump)

J. M. Murray and A. Weber (1974). *Scientific American,* **230**, No.2, 59. (Muscle Contraction)

M. S. Mohan and G. A. Rechnitz. *J. Am. Chem. Soc.,* **92**, 5839 (1970); **94**, 1914 (1972). (Formation Constants of Ca–ATP Complexes)

Experiment 13
The Passage of Metal Ions through Cell Membranes

INTRODUCTION

The chemical structure of cell membranes is not known although their components are phospholipids, proteins and cholesterol as major constituents and sialic acid, phosphatidic acid, glycolipids and inorganic ions (especially calcium) as minor components. In total this means a non-aqueous layer sandwiched between the two aqueous solutions of the intracellular and extracellular fluids. Here we discover a fascinating paradox: on the one hand the membrane is permeable to metal ions of Groups I and II, whereas on the other hand there are definite concentration gradients for these ions across the membrane. For example, the majority of cells have a high internal potassium ion concentration and a low internal sodium ion concentration in contrast to the surrounding fluids where the ratio is reversed, e.g. in red blood cells the $K^+:Na^+$ ratio is 105:10 whereas the ratio is 5:143 in blood plasma. What mechanism produces these essential biological gradients across semipermeable biological membranes? The blunt answer is that we do not know. What we are aware of is that some molecules (called 'carrier molecules') encourage ions to traverse membranes and it is conceivable that similar molecules may well participate in the so-called 'ion pumps' that transport ions to and from cells against a concentration gradient.

Because of our lack of knowledge of *in vivo* membrane processes, of the carrier molecules in membranes and especially because the ions concerned, Na^+, K^+, Mg^{2+}, Ca^{2+}, are not paramagnetic, spectrally detectable, etc. we are not able to provide a crisp introductory bio-inorganic laboratory experiment. We can, however, introduce the student to the fascinating almost unchartered new world of membraneology and describe a demonstration of a carrier molecule at work that poses the more important questions awaiting future research for their solutions.

In 1964, Pressman reported the phenomenon of antibiotic-induced ion transport in mitochondria (these molecules were valinomycin, the gramicidins, macrotetralide actins, etc.). Could similar molecules be involved in ion pumps? A second class of antibiotics (nigericin, monensin, etc.) was found to reverse the ion-transport characteristics of the first group. Both groups of antibiotics were naturally occurring but more recently Pedersen and Lane have synthesised macrocyclic ethers and aminoethers that mimic the ion-pumping behaviour of the antibiotics.

If theories are to be founded upon the membrane-pumping selectivity sequences of these model compounds, simple chemical models of cell membranes are desirable so that selectivity sequences for cation transport into the pseudomembrane (the organic solvent layer) may be established.

PROCEDURE

(a) Partition measurements

The selectivity sequences for simulated transport may be derived using a wide variety of techniques. Salt extraction by antibiotics and other models from aqueous solutions into organic solvents has been widely used to establish selectivities. The experiment involves shaking an organic solvent phase containing the alleged carrier with aqueous solutions of alkali metal salts containing a lipid-soluble, coloured anion. The intensity of colour of the organic phase is then monitored spectrophotometrically to determine the degree of extraction. (Radioisotopes of the cation may also be used as probes, the migration of radiation into the organic phase being the measure of transport.) The extractions of picrate, or dinitrophenolate, salts into n-hexane, dichloromethane or mixtures of solvents are convenient systems to study as these solvents may be taken to represent the dielectric properties of the membrane interior. The results obtained from these experiments can be compared with those designed to show the effects of the carriers on the electric-resistance properties of artificial phospholipid bilayers (*Table VI.13.1*). Compare your spectrophotometric results with *Table VI.13.1*.

Table VI.13.1
Selectivity Sequences from Solvent Extraction Data

Valinomycin	: Rb > K > Cs > Na
Actins	: K > Rb > Cs > Na > Li
Antamanide	: Na > Li~K > Rb~Cs
Nigericin	: K > Rb > Na > Li
Monensin	: Na > K > Rb,Li
Dianemycin	: Na, K, Rb > Li
Dibenzo-18-crown-6	: K > Cs > Na > Li
Dicyclohexyl-18-crown-6	: K > Cs > Na > Li
Benzo-15-crown-5	: Na > K > Cs > Li

(b) U-tube observations

An interesting and effective modification of procedure (a) is the use of U-tubes to provide a transfer simulating system (*Figure VI.13.1*).

A semipermeable barrier of chloroform is interposed between two aqueous solutions and the appropriate antibiotic, etc., is dissolved in the chloroform layer. The metal salts are dissolved in the side-arms and the movement of salt through the barrier is monitored – spectrophotometrically for the picrate ion, by pH meter for protons and by atomic absorption for metal ions. Gently stirring the water layers speeds up the experiment. Selectivity sequences can then be compared for different antibiotics and your results referred to those reported by 'professional' researchers in *Table VI.13.1*.

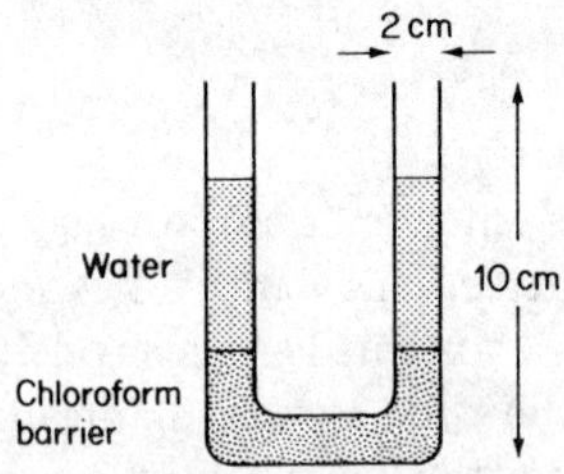

Figure VI.13.1 U-tube experiment using a chloroform layer as a model of a biological membrane

SUGGESTIONS

Either summarise current concepts concerning the 'structures' of cell membranes *or* critically assess mechanisms through which cations can traverse cell membranes.

REFERENCES

D. E. Fenton (1976). *An Introduction to Bio-inorganic Chemistry*, Ed. D. R. Williams, Thomas, Illinois, 281-302
'Symposium on Biological and Artificial Membranes', *Fed. Proc.*, **27**, 1269 (1968)
S. J. Singer and G. L. Nicholson (1971). 'The Fluid Mosaic Model of the Structure of Cell Membranes'. *Science*, **175**, 720
G. Eisenmann, G. Czabo, S. Ciani, S. McLaughlin and S. Krasne (1973). 'Ion Binding and Ion Transport Produced by Neutral Lipid Soluble Molecules'. *Progr. Surface Membrane Science*, **6**, 139
P. B. Chock and E. O. Titus (1973). 'Alkali Metal Ion Transport and Biological Activity'. *Progr. Inorg. Chem.*, **18**, 287
E. -I. Ochiai (1977). *Bio-inorganic Chemistry: An Introduction*, Allyn and Bacon, Boston, Chapter 15

Experiment 14
Stereochemistry of Co(III)(trien)X_2, and Reaction of Coordinated Glycine Esters

INTRODUCTION

Triethylenetetramine (trien) can coordinate to a metal ion in a number of different ways. The possible modes are shown in *Figure VI.14.1* for the case of Co(III) complexes. This sort of structural difference around a metal ion is of importance for a consideration of the catalytic sites of metalloenzymes.

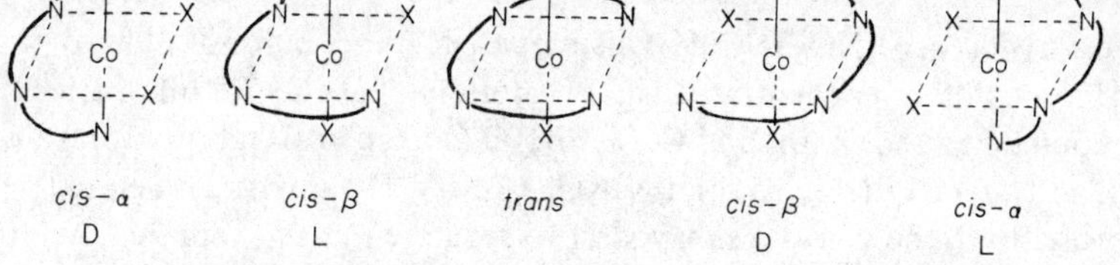

Figure VI.14.1 Geometrical isomers of Co(III)(trien) X_2 complexes

Particularly important in actual living systems is the occurrence of optical isomerism. In many cases only one of the optical isomers is biologically active. In this experiment, however, only geometrical isomers (*cis-α*, *cis-β* and *trans*) will be considered.

Most enzymic reactions may be described as follows:

$$\text{E (enzyme)} + S_1 \text{ (substrate 1)} \rightleftharpoons ES_1 \qquad (14.1)$$

$$ES_1 + S_2 \text{ (substrate 2)} \rightleftharpoons \text{E} + \text{P (product)} \qquad (14.2)$$

In the case of a metalloenzyme with a metal ion M at its catalytic site, the substrate S_1 coordinates to M and is activated or made more labile for the subsequent reaction. In effect, reaction 14.2 is that of the ligand S_1 coordinated on to M. The mechanism of this labilisation, or activation by coordination, is manifested in a number of ways. A frequently utilised mechanism involves the inductive effect exerted by the metal ion; the metal ion (Co^{2+}, Zn^{2+}, Cu^{2+} or Fe^{3+}, etc.) acts as a kind of super acid. In addition to the general inductive effect, the metal ion may exert a specific effect by coordinating a substrate in a specific fashion (e.g. coordination at two *cis*-positions).

Peptide formation or protein synthesis in biological systems is done by a series of enzymes and DNA–RNAs, and there is no known metalloenzyme involved in the process. However, the metabolism of proteins or peptides. (degradation or hydrolysis) may require a metal ion in one step or another, although most peptidases or proteases are simple proteins. Reaction 14.3 which will be tried in this experiment

$$\underset{\substack{|\\ NH_2\,(\text{glyOEt})}}{2CH_2COOEt} + [Co^{III}(\text{trien})X_2] \rightarrow [Co^{III}(\text{trien})\text{glycylglyOEt})]^{3+} \qquad (14.3)$$

therefore is not directly relevant to an actual biological system, although it demonstrates that this peptide formation can be effected readily when the glycine ester coordinates to a metal ion. Under comparable reaction conditions in the absence of the cobalt complex, the glycine ester does not react to form the dipeptide.

STEREOCHEMISTRY

Preparation of *cis*-α-[Co(trien)Cl_2]Cl

First, cis-α[Co(trien)$(NO_2)_2$]Cl.H_2O is prepared. An ice-cold solution of trien.HCl (4.4 ml of concentrated HCl is carefully added to 7.5 ml of trien while cooled in an ice bath) and $CoCl_2.6H_2O$ (11.6 g in 10 ml of water) are mixed, and added quickly to solid $NaNO_2$ (7.3 g). The mixture is aerated vigorously; the yellow crystals may start to separate in a few minutes. Aeration is continued for 1.5 h at 0°C. The product is filtered, washed with ice-cold 1 mol dm^{-3} HCl solution, alcohol, and acetone, and air-dried.

A mixture of the *cis*-α-[Co(trien)$(NO_2)_2$]Cl.H_2O thus prepared (the whole product in 10 ml of water) and excess of concentrated HCl (7 ml) is evaporated on a boiling water bath at 100°C to dryness. The purple product is ground in ethanol, filtered, washed with ethanol and acetone, and air-dried. The product is a mixture of *cis*-α- and *cis*-β-[Co(trien)Cl_2]Cl (*cis*-α predominant). The pure *cis*-α-isomer may be obtained by re-crystallisation twice from boiling 3 mol dm^{-3} HCl aqueous solution. The pure *cis*-α isomer is *not* required, however, for the subsequent preparations.

Preparation of *cis*-β-[Co(trien)Cl_2]Cl.0.5H_2O

A mixture of crude *cis*-α-[Co(trien)Cl_2]Cl (4 g) and Li_2CO_3 (1.2 g) in water (10 ml) is heated on a steam bath with occasional stirring. The violet colour soon changes to red, and the excess Li_2CO_3 is filtered off after 1 h of heating. After cooling to room temperature, 0.2 g of $CaCl_2$ is added to eliminate CO_3^{2-} ion, and the solution is filtered again. From the filtrate, *cis*-[Co(trien)(CO_3)]Cl.1.5H_2O is obtained by carefully adding ethanol. Ethanol is added little by little to the aqueous red solution with brisk swirling until the first permanent turbidity is observed. At this stage leave the turbid solution stoppered in a refrigerator. After 1 day (or 1 week), the red crystals are collected on a sintered glass funnel, washed with ethanol and acetone, and air-dried.

The whole *cis*-[Co(trien)(CO_3)]Cl.1.5H_2O (⩾2 g, predominantly *cis*-β) is dissolved in excess ice-cold concentrated HCl (e.g. 2 g of complex in 5 ml of acid) in a conical flask, which is stoppered and left in a refrigerator. The first two crops (1–2 days or 1 week) are predominantly reddish purple *cis*-β-[Co(trien)-Cl_2]Cl. More product is obtained subsequently if the product is kept to be filtered off and the solution is kept cold.

Preparation of *trans*-[Co(trien)Cl_2]Cl.1.5H_2O

0.5 g of *cis*-β-[Co(trien)Cl_2]Cl is suspended in methanol (40 ml) and refluxed for 30 min. The unreacted violet material is collected and refluxed again in fresh methanol for 30 min. The two methanol filtrates, which are greenish, are combined and flash-evaporated to dryness. To the solid (brownish green) in the flask, a small amount of 6 mol dm^{-3} HCl (3–5 ml) is added. The *cis*-isomers dissolve to form a purple solution. The green crystals, which remain undissolved, are quickly collected on a funnel and air-dried.

SPECTRA

Record IR spectra (Nujol and hexachlorobutadiene mull) and visible spectra of *cis*-α-, *cis*-β-[Co(trien) Cl_2]Cl (in 6 mol dm^{-3} HCl) and *trans*-[Co(trien)Cl_2]Cl (in methanol). Interpret these as far as you can, especially regarding the differences between the isomers.

REACTION OF COORDINATED LIGANDS

Preparation of *cis*-β-[Co(trien)(glycylglyOEt)]$(ClO_4)_3.H_2O$

A mixture of the crude *cis*-α-[Co(trien)Cl_2]Cl complex (0.93 g) glycine ethyl ester hydrochloride (0.84 g) and diethylamine (0.44 g) in 10 ml of dimethyl sulphoxide (DMSO) is heated at 40–50°C with stirring for 2 h. The unreacted violet material is filtered off, and washed with methanol. The methanol wash is combined with the DMSO filtrate, to which acetone is added to precipitate the product. The very hygroscopic greyish-orange precipitate is collected quickly on a funnel. The crystals are dissolved in 3-5 ml of water, and 1 g of solid $NaClO_4$ is added to the solution. The solution in a small Erlenmeyer flask is left open in a refrigerator for 1 day (or 1 week). The orange crystals formed are collected on a funnel, washed with cold water, and ethanol, and air-dried.

Analysis

The compound may be analysed for carbon, hydrogen and nitrogen. The IR spectrum is quite informative in this case for the elucidation of the structure. The visible spectrum may be recorded.

Other experiments which can be performed if time allows

(a) Preparation of *cis*-Co(trien)(glyOEt)Cl]$^{2+}$, and its reaction with another mole of glyOEt to form the complex with coordinated dipeptide.
(b) Saponification of *cis*-[Co(trien)(glycylglyOEt)]$^{3+}$ to form *cis*-[Co(trien)-(glycine)]$^{2+}$.
(c) Aquation reactions of *cis*-α- and *cis*-β-[Co(trien)Cl_2]$^{2+}$.

DISCUSSION

Include some discussion on the stability of the different configurational isomers, on the IR and visible spectral interpretation, and on the mechanism of the dipeptide formation.

REFERENCES

J. P. Collman and E. Kimura (1967). *J. Am. Chem. Soc.*, **89**, 6096
D. A. Buckingham, L. G. Marzilli and A. M. Sargeson (1967). *J. Am. Chem. Soc.*, **89**, 2772, 4539
A. M. Sargeson and G. H. Searle (1967). *Inorg. Chem.*, **6**, 787

Experiment 15
Effects of Axial Ligands on Tetragonal Metal Complexes Related to Haemoglobin, etc.

Table VI.15.1 is a collection of data about the magnetism of some Fe–porphyrin complexes. The porphyrin ligand (*Figure VI.15.1*) involved in these complexes is approximately square planar with four nitrogen atoms to coordinate to iron. *Table VI.15.1* shows that the electronic configuration of these types of complexes greatly depends on the axial ligand, X. Strongly coordinating groups such as CN^- and N_3^- for Fe^{III}; and CN^-, CO and O_2 for Fe^{II} produce a low-spin

Table VI.15.1
Electronic Structure of [Fe(porphyrin)XB]

Complex	X	Magnetic moment (B.M.)	Number of unpaired electrons	Spin state
Fe^{II} haemoglobin		5.44	4	high
	$-O_2$	0	0	low
	$-CN^-$	0	0	low
	-CO	0	0	low
Fe^{III} methaemoglobin	$-H_2O$	5.73	5	high
	$-F^-$	5.77	5	high
	$-N_3^-$	3.30	–	mixture of high and low
	$-CN^-$	1.96	1	low
Fe^{III} peroxidase	$-F^-$	5.90	5	high
	$-H_2O$	5.48	5	high
	$-OH^-$	2.66	–	mixture of high and low
	$-CN^-$	2.67	–	mixture of high and low

state for these series of compounds (d^5 and d^6, respectively, in a tetragonal field), while weaker ligands such as H_2O and F^- produce high-spin states.

This experiment is designed to show this sort of axial ligand effect on the electronic structure of tetragonal complexes. The compounds are $Ni^{II}(Et_2en)_2X_2$. Ni^{II} has a d^8 configuration (Et_2en = *N,N*-diethylethylenediamine). In an

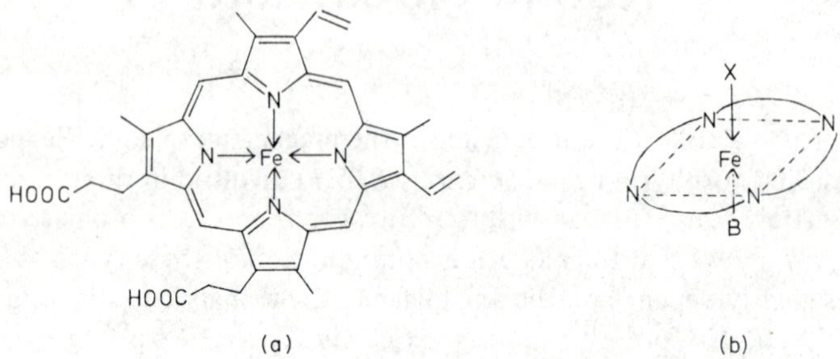

Figure VI.15.1 (a) Protoporphyrin-IX – the most common porphyrin in biology – and (b) a schematic structure of [Fe(porphyrin) XB] where B = imidazole, etc.

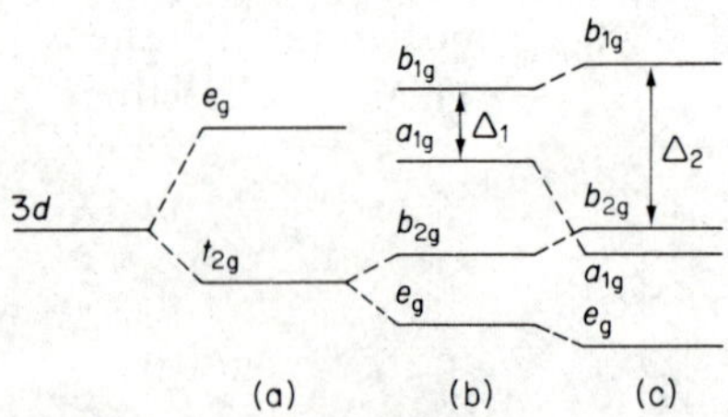

Figure VI.15.2 Schematic diagram of energy levels of 3*d* in (a) octahedral, (b) weak tetragonal distortion from octahedral (tetragonal), and (c) strong tetragonal distortion (square planar)

octahedral field, the 3*d* level splits into e_g and t_{2g} levels. With a weak tetragonal distortion, t_{2g} splits into e_g (d_{yz}, d_{zx}) and $b_{2g}(d_{xy})$, while e_g splits into $a_{1g}(d_{z^2})$ and b_{1g} ($d_{x^2-y^2}$). The order of these levels (orbitals) and the energy difference between them depend on the strength of this tetragonal field, as shown in *Figure VI.15.2.* If Δ_1 is not very large, i.e. the field is weakly tetragonally distorted, the ground electronic configuration would be $(e_g)^4$ $(b_{2g})^2(a_{1g})^1(b_{1g})^1$ which is paramagnetic with two unpaired electrons, i.e. $^3B_{1g}$. On the other hand, for such a complex where Δ_2 is large and the d_{z^2} level lies fairly low, the ground electronic configuration would be $(e_g)^4(a_{1g})^2(b_{2g})^2$ which is diamagnetic, i.e. $^1A_{1g}$. The position of the d_{z^2} level depends greatly on the axial ligand, X, because the d_{z^2} orbital is directed towards X. If X is a strongly coordinating ligand, the interaction would raise the energy of the d_{z^2} level (Δ_1 = small, weak tetragonal distortion from octahedral). On the other hand, if the interaction between d_{z^2} and X is small (poorly coordinating X), the d_{z^2} energy level is lower (or is not destabilised).

Thus, a strong field X favours the paramagnetic state while a weak field X favours the diamagnetic state. These two types of compounds, $Ni(Et_2en)_2X_2$, distinguished by magnetic properties have also quite distinct and different colours. The paramagnetic ones are blue to violet while the diamagnetic ones are yellow-orange-red. This difference in colour may also be explained in terms of the electronic configuration discussed above.

PREPARATION OF COMPOUNDS

$Ni(Et_2en)_2Cl_2.2H_2O$

Add 2.32 g of Et_2en to an ethanolic solution (70 ml) of 2.38 g of $NiCl_2.6H_2O$ with stirring at room temperature. In 10–20 min, light-blue crystals form. Collect them on a sintered glass funnel, wash once (quickly) with 10 ml of ethanol and air-dry by suction.

$Ni(Et_2en)_2(NCS)_2$

Prepare ethanol solutions of $Ni(NO_3)_2.6H_2O$ (2.91 g in 40 ml) and NaNCS (1.62 g in 30 ml). Mix them with stirring. In 10–15 min filter off the white precipitate formed. Add 4 ml of ethanol and then 2.32 g of Et_2en to the green filtrate with stirring. Collect the violet crystals which form in a few minutes on a sintered glass funnel, wash with 20 ml of ethanol and air-dry.

$Ni(Et_2en)_2(NO_3)_2$

Add 2.32 g of Et_2en to an ethanol solution (50 ml) of 2.91 g of $Ni(NO_3)_2$-$.6H_2O$ with stirring at room temperature. In 20 min orange crystals form. Collect them on a sintered glass funnel, wash with 20 ml of ethanol and air-dry.

$Ni(Et_2en)_2I_2$

Prepare ethanolic solutions of $Ni(NO_3)_2.6H_2O$ (2.91 g in 40 ml) and NaI (3.00 g in 30 ml). Mix them with stirring. In a few minutes, filter off the white precipitate formed, and then add 5 ml more of ethanol to the filtrate. Add 2.32 g of Et_2en to the filtrate. Collect the red-orange crystals, which immediately form, on a sintered glass funnel, wash with 20 ml of ethanol and air-dry.

MEASUREMENT OF MAGNETIC SUSCEPTIBILITY

Measure the magnetic susceptibility, estimate the effective magnetic moment, and discuss the electronic configuration for the compounds.

PROCEDURE

Record the following items.

(i) Weight of the sample tube (dry and clean) in (I_i) and out of (I_o) the magnetic field:

$$\Delta r = I_i - I_o$$

(ii) Weight of the standard compound, Hg[Co(SCN)$_4$] (in the same sample tube, dry and clean) in (II_i) and out of (II_o) the magnetic field:

$$\Delta W_2 = II_i - II_o, W_2 = II_o - I_o$$

(iii) Weight of the sample compound (in the same sample tube, dry and clean) in (III_i) and out of (III_o) the magnetic field:

$$\Delta W_3 = III_i - III_o, W_3 = III_o - I_o$$

Proper packing of the solid sample in the tube is one of the important factors in getting reliable data. Pulverise the solid sample finely (not too much) and pack it in 10–15 aliquots of the total amount needed. After addition of each aliquot into the tube, lightly tap the bottom of the tube on a bench 30–40 times. Pack the solid sample to a certain height (5–7 cm). This height must be kept the same (as exact as possible) for a series of measurements.

There are a number of susceptibilities to be defined. χ_v is volume-susceptibility and is, in fact, measured directly by this method (Gouy method); χ_g is susceptibility per gram of sample and χ_M is molar susceptibility. They are related by the following equations:

$$\chi_g = \chi_v/\rho \ (\rho = \text{density of solid sample}) \qquad (15.1)$$

$$\chi_M = M\chi_g \ (M = \text{molecular weight of the compound}) \qquad (15.2)$$

χ_v is related to the magnetic field strength (H) etc. in the following way:

$$\chi_v\,(\text{sample}) - \chi_v\,(\text{air}) = \frac{2\Delta W}{aH^2} \qquad (15.3)$$

χ_v(air) is usually negligible, a is a constant, and ΔW is defined above. From equation 15.1, $\chi_v = \chi_g\rho$ and ρ are given by W/V (W = weight of the sample as defined above, and V = volume of the solid sample), therefore

$$\chi_g = \frac{V}{W}\chi_v = \frac{2V}{aH^2}\cdot\frac{\Delta W}{W} = B\frac{\Delta W}{W}\left(\text{where } B = \frac{2V}{aH^2}\right) \qquad (15.4)$$

ΔW must be corrected for the contribution of the tube (Δr). B is constant under the condition of constant field strength (H) and sample volume (V). Constant volume is ensured by packing the tube to a constant height. For Hg(Co(NCS)$_4$], χ_g is measured to be 16.44 x 10^{-6} c.g.s. ≡ 206.6 x 10^{-12} M^3 kg^{-1} (in S.I. units). From ΔW_2 (corrected) and the W_2 value for the standard, calculate the

B value. Using this B value and your data for ΔW_3, calculate χ_g for your compound. Convert χ_g to χ_M.χ_M of a complex consists of the summation of χ_M values for the central ion, the coordinated ligand and the other species present (such as solvent molecules, associated anions or cations); thus

$$\chi_M \text{ (complex)} = \chi_M{}^1 \text{ (metal ions)} + \chi'_M \text{ (ligands)} + \chi_M \text{ (other species)} \qquad (15.5)$$

χ'_M for ligands and other species are given below.

Estimate χ'_M (central metal ion). χ'_M is related to the effective magnetic moment μ_{eff} in the following ways:

$$\left.\begin{aligned} \mu_{eff} &= 2.83(\chi'_M \cdot T)^{1/2} \text{ B.M. } (\chi'_M \text{ in c.g.s. units}) \\ \text{or} \quad &= 7.40 \times 10^{-21} (\chi'_M \cdot T)^{1/2} \text{ A m}^2 \ (\chi' \text{ in S.I. units}) \\ \text{or} \quad &= 798(\chi'_M \cdot T)^{1/2} \text{ B.M. } (\chi'_M \text{ in S.I. units}) \end{aligned}\right\} \qquad (15.6)$$

In the above equations, T is the absolute temperature of measurement. μ_{eff}, in turn, is related to the number of unpaired electrons, n, and is expressed by the following equation if the magnetic contribution from the orbital angular momentum of the electrons can be neglected:

$$\mu_{eff} = (n(n+2))^{1/2} \qquad (15.7)$$

Calculate μ_{eff} and estimate n to the nearest integral value. The molar susceptibilities of the ligands are given below.

χ_{Et_2en}	$= -101 \times 10^{-6}$ c.g.s. ($cm^3\ mol^{-1}$)	=	$-1269 \times 10^{-12}\ m^3\ mol^{-1}$	(S.I. units)
χ_{Cl^-}	$= -25 \times 10^{-6}$ c.g.s.	=	$-314 \times 10^{-12}\ m^3\ mol^{-1}$	
χ_{H_2O}	$= -13 \times 10^{-6}$ c.g.s.	=	$-163 \times 10^{-12}\ m^3\ mol^{-1}$	
χ_{NCS^-}	$= -33 \times 10^{-6}$ c.g.s.	=	$-415 \times 10^{-12}\ m^3\ mol^{-1}$	
χ_{I^-}	$= -51 \times 10^{-6}$ c.g.s.	=	$-641 \times 10^{-12}\ m^3\ mol^{-1}$	
$\chi_{NO_3^-}$	$= -19 \times 10^{-6}$ c.g.s.	=	$-239 \times 10^{-12}\ m^3\ mol^{-1}$	

OTHER EXPERIMENTS WHICH CAN BE PERFORMED IF TIME ALLOWS

Electronic spectra

Record the (visible) spectra of the compounds obtained by the Nujol mull method for $Ni(Et_2en)_2(NO_3)_2$ and $Ni(Et_2en)_2I_2$, and solution method (10^{-2} mol dm^{-3}) for $Ni(Et_2en)_2Cl_2.2H_2O$ (methanol) and $Ni(Et_2en)_2(NCS)_2$ (in CH_2Cl_2). Discuss the spectra in terms of electronic structure.

Electric conductance

In nitrobenzene: 10^{-3}–0.4×10^{-3} mol dm^{-3}

Et_2en = N, N-diethylethylenediamine.

Preparation of other compounds

$Ni(Et_2en)X_2$ with X = BF_4^-, NO_2^-, Br^-, etc.

REFERENCES

D. M. L. Goodgame and L. M. Venanzi (1963). J. Chem. Soc., London, 616-627

J. Lewis and R. G. Wilkins (1960). *Modern Coordination Chemistry*, Interscience, 403

P. W. Selwood (1956). *Magnetochemistry*, Interscience, Chapters 5 and 6

For magnetic properties of hemo proteins: M. Weissbluth (1967). *Structure and Bonding*, **2**, 1-125

Experiment 16
Interaction of Pyrophosphate with Metal Ions

INTRODUCTION

Phosphorus is one of the essential elements for biological systems, and exists mostly either in the form of inorganic phosphate, or in the form of nucleotide or polynucleotide. One of the vital molecules for all the living systems is ATP, adenosine triphosphate, a nucleotide which consists of adenine, a carbohydrate residue and a triphosphate group. The hydrolysis of ATP to ADP, adenosine diphosphate, and a phosphate provides the energy required in biological systems: muscle contraction, chemical energy for synthesis, and so on.

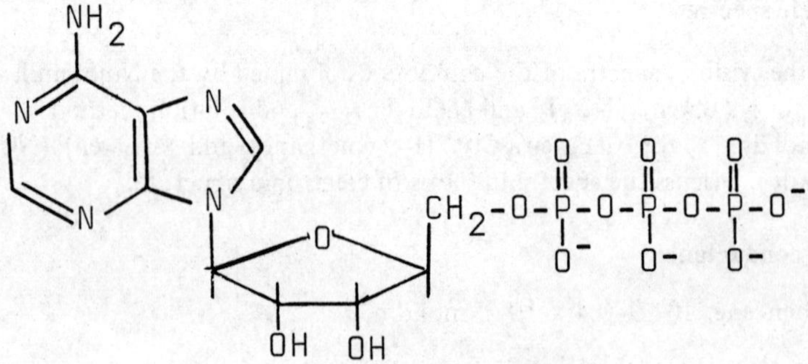

The enzymes which catalyse this hydrolysis and/or phosphate transfer reactions require Mg^{2+}, K^+, Na^+, Ca^{2+} or Mn^{2+} for activity. ATP itself, being highly charged, exists mostly chelated to Mg^{2+} in cells.

Since phosphate is vital to the living systems, the dumping of a large quantity of phosphate in the environment, particularly the hydrosphere, causes the so-called 'eutrophication'. One of the sources of such phosphates released to the environment is detergent, which contains phosphate derivatives (mostly in the form of polyphosphates). The reason why the phosphates are used as an ingredient in detergents is that they are capable of chelating metal ions in filth or stain and making them soluble, and also of softening hard water.

This experiment investigates the types of chelates formed in the pyrophosphate-metal ion systems. The pyrophosphate, $P_2O_7^{4-}$, has the following structure and may serve as a model for the phosphate part of ATP and ADP.

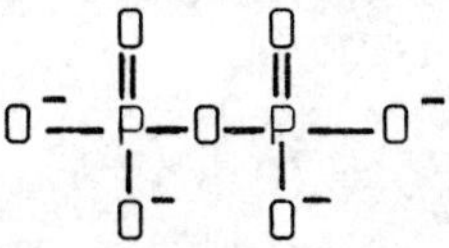

THEORY OF CONDUCTOMETRIC TITRATION TECHNIQUE

The conductometric titration technique is used to determine the types of the chelates. A solution of sodium pyrophosphate is titrated with a metal salt solution, and the specific conductance of the solution is measured at each addition of the titrant, by a conductance cell-bridge. In a dilute solution the specific conductance can be expressed as a sum of the contributions from all ionic species involved:

$$L \text{ (specific conductance)} = \sum_i c_i \lambda_i \qquad (16.1)$$

Here c_i is the molar concentration of an ionic species, i, and λ_i its molar conductance; λ_i may be expressed by the equivalent conductance, λ_i^0, when the ion carries a charge $\pm n$, as follows:

$$\lambda_i = n\lambda_i^0 \qquad (16.2)$$

Suppose that a solution containing $4A^+$ and B^{4-} is titrated with a solution containing C^{2+} and $2D^-$. The reaction would then be:

$$m(4A^+ + B^{4-}) + n(C^{2+} + 2D^-) \rightleftharpoons (C_nB_m)^{2n-4m} + 4mA^+ + 2nD^- \qquad (16.3)$$

$$(C_nB_m)^{2n-4m} + n'C^{2+} \rightleftharpoons (C_{n+n'}B_m)^{2n+2n'-4m} \qquad (16.4a)$$

Suppose also that the maximum ratio of m/n is 2 and that the $[CD_2]$ added is less than $1/2\,[A_4B]$. The entities such as A^+, B^{4-}, D^- and $(CB_2)^{6-}$ are now present in the solution. Let a be the initial concentration of A_4B and x be the concentration of CD_2 added by this time. Then the concentrations of the species are $[CB_4^{6-}] = x$, $[B^{4-}] = a-2x$, $[A^+] = 4a$ and $[D^-] = 2x$. The molar conductance of the species are denoted as $\lambda_{CB_2} = 6\lambda^0_{CB_2} \equiv 6\lambda^0_1$ and $\lambda_{B^{4-}} = 4\lambda^0_B$. Then the conductance of this dilute solution should be $L = 6\lambda^0_1 x + 4\lambda^0_B(a-2x) + 4a\lambda^0_A + 2x\lambda^0_D = (6\lambda^0_1 + 2\lambda^0_D - 8\lambda^0_B)x + 4(\lambda^0_B + \lambda^0_A)a$. If all the entities behave ideally and can be regarded as sufficiently small balls, then $\lambda^0_1 = \lambda^0_D = \lambda^0_B = \lambda^0_A$, and hence, $L = 0x + 4(\lambda^0_B + \lambda^0_A)a$ = constant. That is, if one plots L against x, one would obtain a straight line parallel to the x-axis. However, this is not the case in practice; that is λ^0_i values are not necessarily the same. The λ^0 value for a large complicated molecule is usually smaller than that for a smaller and simpler entity. Therefore, L can be expressed by $L = u_1x + v_1$ ($u_1 \neq 0$ but small).

Next we suppose that

$$(CB_2)^{6-} + C^{2+} \rightleftharpoons (C_2B_2)^{4-} \qquad (16.4b)$$

and consider the range of $a/2 \leqslant x < a$. In the titration solution we have now A^+, D^-, $(CB_2)^{6-}$ and $(C_2B_2)^{4-}$. The concentrations are $[A^+] = 4a$, $[D^-] = 2x$, $[CB_2^{6-}] = a-x$, and $[C_2B_2^{4-}] = x-a/2$. Therefore, $L = 4a\lambda^0_A + 2x\lambda^0_D + 6\lambda^0_1(a-x) + 4\lambda^0_2(x-a/2) = (4\lambda^0_A + 6\lambda^0_1 - 2\lambda^0_2)a + (2\lambda^0_D + 4\lambda^0_2 - 6\lambda^0_1)x = u_2x + v_2$ ($u_1 \neq u_2 \neq 0$ but small; $\lambda^0_2 = \lambda^0_{C_2B_2}$). If we plot the experimental value of L against x, we would expect such a graph as shown in *Figure VI.16.1*. The inflection points in the conductimetric titration can thus be used to determine the compositions of the complexes species C_nB_m.

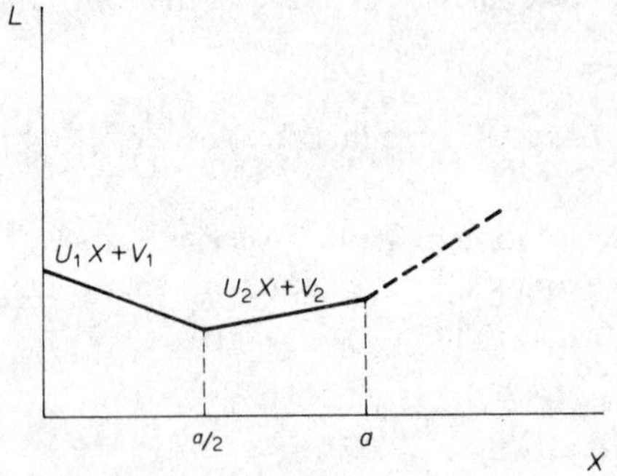

Figure VI.16.1 Plot of L versus X as the titration proceeds

PROCEDURE

Prepare the following solutions:
100 ml of 0.100 mol dm^{-3} solution of $Na_4P_2O_7.10H_2O$;
50 ml of 0.100 mol dm^{-3} solution of $MnSO_4.H_2O$;

50 ml of 0.100 mol dm^{-3} solution of $MgSO_4$;
50 ml of 0.100 mol dm^{-3} solution of $CaCl_2.6H_2O$;
50 ml of 0.100 mol dm^{-3} solution of $CuSO_4.5H_2O$.

Take 2 ml of the 0.1 mol dm^{-3} solution of $Na_4P_2O_7.10H_2O$, and dilute it to 500 ml. Transfer this solution into a 1000-ml beaker and insert a conductance cell. Titrate this solution with the metal salt solution from a 10-ml burette. Stir the solution continuously except for the time of reading the conductance. Read the conductance (or resistance) after every 0.20-ml addition of the titrant. Plot the conductances against the added volume of the titrant, which may be approximated to the increase of the concentration of the titrating species (since the volume change up to 2% is negligible).

Interpret the results, and discuss the types of the chelates formed and the possible structures of the chelates.

REFERENCE

L. B. Rogers and C. A. Reynolds (1949). *J. Am. Chem. Soc.*, **71**, 2081; the data should not be relied upon

Experiment 17
Metal Ion–Protein Interactions

INTRODUCTION

The possible coordination sites for metal ions in proteins are -COO^-, –ArOH (tyrosine), -OH (serine, threonine), –S^- (cysteine), -NH_2, -CONH-, and imidazole (histidine). There are also a few coordinating groups other than amino-acid residues in certain proteins; e.g. heme group. The different combinations of coordination sites, coordination structure, bond nature, and metal ions, create a wide range of activities of metalloenzymes. The elucidation of the mechanism of metalloenzyme reactions thus requires the identification of the nature of the metal-protein system as outlined above (coordinating group, etc.). The -S^- residue of cysteine is particularly important, as it serves as a catalytic site in many enzymes, and is readily susceptible to reaction with heavy metals such as Pb and Hg. This forms the basis for the mechanism of toxicity of the heavy metals at the atomic and molecular level. Imidazole very often acts as a general acid-base catalytic centre in enzymes, and also strongly coordinates to certain metal ions.

There are a great number of copper-containing enzymes known: a few of these are ceruloplasmin, hemocyanin, amine oxidase and laccase. Serum albumin, a protein which has one SH group (cysteine), is considered to be the Cu-transporting protein in blood.

In this experiment, the system Cu^{2+} – bovine serum albumin (BSA, molecular weight = 65 000) is studied spectrophotometrically.

PROCEDURE

Prepare the following solutions:
25 ml of 2% BSA solution in 0.33 mol dm^{-3} aqueous $NaOOCCH_3$ (0.5 g in 25 ml);
50 ml of *ca.* 1.5 x 10^{-2} mol dm^{-3} aqueous solution of $CuSO_4.5H_2O$;
50 ml of *ca.* 1.0 x 10^{-3} aqueous solution of $Hg(ClO_4)_2$ (or $HgCl_2$).

In the preparation of BSA solutions, the undissolved material (if any) may be centrifuged off. Using these solutions, make up the solutions listed below and record their visible spectra (810–600 nm and 480–330 nm) in the optical cell provided.

Solution No.	BSA solution (ml)	Cu solution (ml)	Hg solution (ml)	Water (ml)
1	3.00	1.00	0	1.00
2	3.00	0.80	0	1.20
3	3.00	1.00	1.00	0
4	0	1.00	0	4.00
5	3.00	0	0	2.00
1′, 3′, 4′	10 mg Na_2EDTA (solid) + 1, 3, 4 solutions			

The solutions 1′, 3′, and 4′ do not have to be prepared separately. Simply add the solid Na_2EDTA to the respective solutions in the cell after recording their spectra.

Interpret your spectra and discuss the results, especially regarding the binding sites for copper.

REFERENCES

H. A. Fiess and I. M. Klotz (1952). *J. Am. Chem. Soc.*, **74**, 887
I. M. Klotz, J. M. Urquhart and H. A. Fiess (1952). *J. Am. Chem. Soc.*, **74**, 5537
B. L. Vallee and W. E. C. Wacker (1970). *The Proteins, Vol. V*, Academic Press (a general reference)

Part C

Enzymes, Proteins and Model Studies

Chapter VII

Introduction

About one-third of all known enzymes require metal ions in one way or another. Some enzymes have metal ions strongly bound to their own protein structures; these are called metalloenzymes. Others require added metal ions as a co-factor. The studies on metalloenzymes and metal-activated enzymes constitute a major portion of bio-inorganic chemistry. Proteins containing metal ions as an active factor are called metalloproteins. They function mostly as oxygen carriers, electron carriers and metal carriers. The studies on metalloproteins are not less important than those on metalloenzymes. Some of the important metalloproteins, metalloenzymes and metal-activated enzymes are listed in *Table VII.1.*

Table VII.1
Examples of Metalloenzymes, Metal-activated Enzymes and Metalloproteins

Fe:	haem prosthetic group: myoglobin, peroxidase, cytochrome P-450, tryptophan dioxygenase, cytochrome *c*
Fe:	non-haem : pyrocatechase, haemerythrin, ferredoxin, transferrin
Cu:	tyrosinase, laccase, superoxide dismutase, haemocyanin, plastocyanin, azurin
Co:	corrin co-enzyme (B_{12} co-enzyme): glutamate mutase, dioldehydrase
Zn:	carbonic anhydrase, carboxypeptidase, alcohol dehydrogenase
Mg:	phosphotransferases (kinases) and phosphohydrases (ATPase etc.)
K:	pyruvate phosphokinase, Na/K-specific ATPase
Na:	Na/K-specific membrane ATPase

Biochemists naturally study the actual biological material itself but inorganic chemists find it also meaningful to study model systems for metalloproteins and metalloenzymes. In the present part (Part C), we study representative systems of metalloproteins, metalloenzymes and their model systems.

METHODS OF STUDY

In research, at least, one has to start from the original biological material; sacrificing animals, extracting proteins and enzymes from tissues or micro-organisms, purifying the proteins and the enzymes, assaying their biological activities, studying their physicochemical properties, and finally investigating their mechanisms of activity. The early portion of this scheme of study will not be covered in this book. Many excellent laboratory manuals are available for the introduction to this field, i.e. biochemistry laboratory manuals. For the purpose of research, one of the best sources of information is *Methods in Enzymology* (Academic Press).

The studies of the physicochemical properties and the mechanisms of metalloenzymes and metalloproteins seem to be proper to bio-inorganic chemistry. In this chapter, we will be using commercially available purified material.

The methods available for investigating the physicochemical properties of coordination compounds have been outlined in the previous part (Part B). The techniques applicable to the studies of enzymes and proteins are essentially the same as those applied to simpler coordination compounds, except for some difficulties involved in handling biological material. The difficulties are as follows. (i) Scarcity of material. Only a limited amount of material is available in many cases. Therefore, one must be very careful in handling them; do not waste them. Also, one must devise systems in which one could get enough information with a small amount of material; i.e. a high sensitivity in the instruments and the methods is the most important requirement. (ii) Vulnerability of biological material. The materials are very vulnerable to many factors; they will lose activities quite easily. Some of the important factors to which enzymes and proteins are susceptible are temperature, pH and ionic strength of the solution, and consumption by microorganisms.

There is no general methodology to devise models for enzymes and proteins. The minimum requirement for a model is its ability to mimic the biological activity. This does not necessarily guarantee that the mechanism of the activity of the model system is the same as that of the native biological system. Unfortunately, the model systems devised so far, in most cases, mimic only partially the native systems. In the following experiments, some prominent model systems are studies. It is quite important to ponder upon the validity of the model systems in studying those experiments.

ENZYME KINETICS AND MECHANISMS

One of the simplest ways to describe an enzymatic reaction mechanism is shown by the following equation:

$$\text{E (enzyme)} + \text{S (substrate)} \underset{k_{-1}}{\overset{k_1}{\rightleftharpoons}} \text{ES} \qquad \text{(VII.1)}$$

$$\mathrm{ES} \xrightarrow{k_2} \mathrm{E} + \mathrm{P}\ (\text{product}) \qquad \text{(VII.2)}$$

If one assumes a steady state condition, i.e. $d[ES]/dt = 0$, then

$$d[ES]/dt = k_1[E][S] - (k_{-1} + k_2)[ES] = 0 \qquad \text{(VII.3)}$$

that is

$$[ES] = k_1[E][S]/(k_{-1} + k_2) \qquad \text{(VII.4)}$$

The total concentration of the enzyme $[E]_0$ is given by

$$[E]_0 = [E] + [ES] \qquad \text{(VII.5)}$$

therefore

$$[E] = (k_{-1} + k_2)[E]_0/((k_{-1} + k_2) + k_1[S]) \qquad \text{(VII.6)}$$

The overall reaction rate is then given by

$$v = -d[S]/dt = d[P]/dt = k_2[ES] \qquad \text{(VII.7)}$$

Equation VII.8 can be written as follows, by use of VII.4 and VII.7

$$v = k_2[E]_0[S]/((k_{-1} + k_2)/k_1 + [S]) \qquad \text{(VII.8)}$$

The parameter $(k_{-1} + k_2)/k_1$ is called the 'Michaelis–Menten' constant, K_m. When [S] is very large, the equation VII.8 becomes

$$v = v_{max} = k_2[E]_0 \qquad \text{(VII.9)}$$

Using the parameters v_{max} and K_m, the equation is rewritten as follows:

$$v = v_{max}[S]/(K_m + [S]) \qquad \text{(VII.10)}$$

This equation is called the 'Michaelis–Menten' equation. If one takes the reciprocal of both sides of the equation one gets:

$$1/v = 1/v_{max} + K_m/v_{max}[S] \qquad \text{(VII.11)}$$

Therefore, if one plots $1/v$ versus $1/[S]$, one would obtain a straight line, and be able to determine the parameters, v_{max} and K_m/v_{max}, from the intercept and the slope of the line, respectively. This is called a 'Linewever–Burk' plot. As implied in the equations above, $v_{max}/[E]_0$ (i.e. k_2) represents the intrinsic activity of the enzyme, and K_m is a measure of the strength by which the enzyme binds the substrate. If $k_{-1} \gg k_2$, $K_m = k_{-1}/k_1$; i.e. K_m^{-1} is equal to the equilibrium constant of reaction VII.1.

Remarkably, many enzymatic reaction rates can be approximated by the Michaelis–Menten equation VII.10. This implies that the mechanism (VII.1 and VII.2) could hold essentially true for many such enzymatic reactions. The reaction VII.1 can be described as follows. The substrate binds to the active site of the enzyme and is activated. The substrate thus activated then reacts (perhaps with another reactant in most cases) to form the product. Bio-inorganic chemistry is concerned with the structure of the active site and how the substrate binds to the active site.

EQUILIBRIUM MODELS

The total metallic content of a biological fluid or sample of tissue is made up of metalloenzymes (as discussed in the Introduction to this chapter), protein–metal ion labile complexes, low-molecular-weight metal complexes and aquated metal ions. These different states are interrelated by the following facts: (i) biological solutions tend to be very close to equilibrium in order to ensure the efficiency of reactions; (ii) the rate of attainment of equilibrium is accelerated by the presence of metalloenzymes; (iii) because of the difficulties associated with analysing the concentrations of many thousands of extremely dilute low-molecular-weight complexes co-existing all in the same solution, computer simulation models are the sole means of studying such phenomena. These have many conceptual factors in common with the enzyme model studies just outlined.

REFERENCES

B. L. Vallee and W. E. C. Wacker (1969). *The Proteins, Vol.5, Metallo-Proteins,* Academic Press

Inorganic Biochemistry, Ed. G. L. Eichhorn, Elsevier, Amsterdam (1973)

E.-I. Ochiai (1977). *Bio-inorganic Chemistry, an Introduction,* Allyn and Bacon, Boston

Methods in Enzymology, Eds. S. P. Colowick and N. O. Kaplan, Academic Press, New York (1955–78)

An Introduction to Bio-Inorganic Chemistry, Ed. D. R. Williams, Thomas, Illinois (1976)

Chapter VIII

Laboratory Experiments

Experiment 18
Haemoglobin – Oxygenation, and Magnetic and Spectroscopic Studies

INTRODUCTION

Haemoglobin is the oxygen carrier in the blood of many animals. The molecule consists of four subunits; in human adults, the subunits are two α-chains and two β-chains. Each subunit has a haem group embedded in it and approximately 150 amino-acid residues. The iron of the haem group, i.e. Fe–protoporphyrin-IX, is bound with a histidine residue in the polypeptide. The molecular weight of the haemoglobin is 64 500. Myoglobin, which is the oxygen-storage protein in muscle and other organs, in contrast, consists of a single chain that is very similar to the α- or β-chain of haemoglobin. The difference in the composition of the molecules makes a remarkable contrast in the oxygen-binding property of the two compounds. The curves of oxygen uptake versus oxygen pressure are shown in *Figure VIII.18.1.* The curve for myoglobin is a normal hyperbola whereas that for haemoglobin is sigmoidal.

The oxygen-uptake equilibrium for myoglobin may be simply described by

$$Mb + O_2 \overset{K}{\rightleftharpoons} MbO_2 \tag{18.1}$$

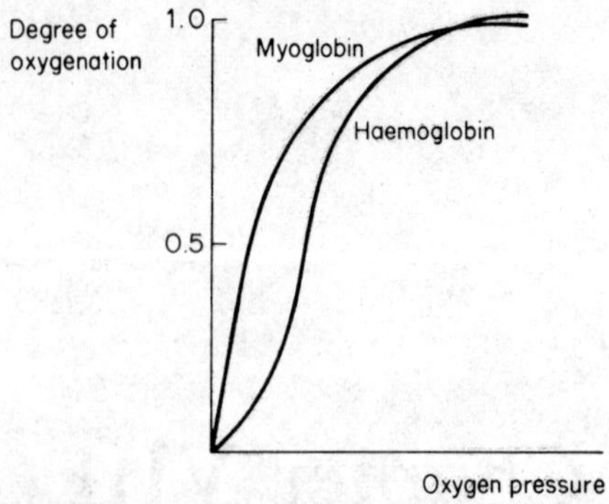

Figure VIII.18.1 Oxygenation curve for haemoglobin and myoglobin

Since haemoglobin has four oxygen-binding sites, the oxygen-binding equilibrium can be described by

$$\left.\begin{array}{llcl} \mathrm{Hb} & +\mathrm{O_2} & \overset{K_1}{\rightleftharpoons} & \mathrm{Hb(O_2)_1} \\ \mathrm{Hb(O_2)_1} & +\mathrm{O_2} & \overset{K_2}{\rightleftharpoons} & \mathrm{Hb(O_2)_2} \\ \mathrm{Hb(O_2)_2} & +\mathrm{O_2} & \overset{K_3}{\rightleftharpoons} & \mathrm{Hb(O_2)_3} \\ \mathrm{Hb(O_2)_3} & +\mathrm{O_2} & \overset{K_4}{\rightleftharpoons} & \mathrm{Hb(O_2)_3} \end{array}\right\} \quad (18.2)$$

The oxygenation fraction y is then given by:

$$y = \frac{(\mathrm{O_2})_{\mathrm{bound}}}{(\mathrm{Hb})_{\mathrm{total}}} = \frac{\mathrm{p}K_1 + 2\mathrm{p}^2K_1K_2 + 3\mathrm{p}^3K_1K_2K_3 + 4\mathrm{p}^4K_1K_2K_3K_4}{4(1 + \mathrm{p}K_1 + p^2K_1K_2 + p^3K_1K_2K_3 + \mathrm{p}^4K_1K_2K_3K_4)} \quad (18.3)$$

This equation is sometimes called the 'Adair equation'. If $K_1 = K_2 = K_3 = K_4 = K$, the saturation factor y can be expressed by the following equation:

$$y = \mathrm{p}K/(1 + \mathrm{p}K) \quad (18.4)$$

This equation is obtained by taking the statistical weight of each species ($\mathrm{Hb(O_2)}_i$) into consideration. This should then give a hyperbolic curve similar to the oxygenation curve of myoglobin. This is equivalent to the Adair equation with the assumption that

$$K_i = K(n - i + 1)/i \quad (18.5)$$

In the present case,

$$K_1{:}K_2{:}K_3{:}K_4 = 4{:}3/2{:}2/3{:}1/4 = 1{:}0.375{:}0.167{:}0.0625 \quad (18.6)$$

This is a result of the assumption that all the haem groups have the same affinity for oxygen, i.e. K, and that the K_i values are determined solely by the statistical weight. The failure of both equations 18.4 and 18.5 to explain the experimental oxygenation curve of haemoglobin indicates that the oxygen affinities of the four haem groups in haemoglobin are not identical and not determined solely by the statistical weight. This means that there is an interaction between haem groups (called 'haem–haem interaction'). The oxygenation curve of sheep haemoglobin has been shown to fit with the relative K values of $K_1:K_2:K_3:K_4$ = 1:1.76:1.31:17.7. This should be compared to the values in equation 18.6.

The haem group of haemoglobin can bind O_2 in the state of Fe(II). CO and NO can bind with the Fe(II) of haemoglobin and myoglobin much more strongly than O_2 does. Fe(II) in haemoglobin and myoglobin can be oxidised to Fe(III). Haemoglobin and myoglobin with Fe(III) are called methaemoglobin and metmyoglobin, respectively. Methaemoglobin and metmyoglobin cannot bind O_2 or CO. Fe(III) in methaemoglobin (and metmyoglobin) bind with a variety of ligands such as OH_2, OH^-, F^-, N_3^-, CN^-, etc. Fe(II) and Fe(III) can take both high-spin and low-spin states. In the case of high-spin Fe(II)–porphyrin and Fe(III)–porphyrin complexes, the iron atom does not lie in the porphyrin plane but a little above it. Thus the iron atom is subject to a lesser ligand field. In the low-spin Fe(II)–porphyrin and Fe(III)–porphyrin complexes, the iron atom is found to be in the porphyrin plane. The conversion of the high-spin to the low-spin state seems to be dependent on the sixth axial ligand (the fifth ligand is the histidine nitrogen). The Fe(II) in the deoxyhaemoglobin and the deoxymyoglobin is in the high-spin state $S = 2$, while the oxyhaemoglobin, carboxyhaemoglobin, and their myoglobin analogues are in the low-spin state, i.e. diamagnetic state ($S = 0$). The Fe(III) derivatives are usually a mixture of the high-spin ($S = 5/2$) and the low-spin ($S = 1/2$) forms; that is

$$\underset{\text{high-spin}}{\text{XFe(III)Hb(or Mb)}} \rightleftharpoons \underset{\text{low-spin}}{\text{XFe(III)Hb(or Mb)}} \tag{18.7}$$

In the case of the complexes with such a weak axial ligand as H_2O or F^-, the high-spin state constitutes the ground state. For the complexes with a stronger axial ligand such as N_3^-, or CN^-, the low-spin state becomes the ground state. The spin-only values of magnetic moment for these spin states are $\mu = 4.90$ B.M. (for $S = 2$), $\mu = 5.92$ B.M. ($S = 5/2$) and $\mu = 1.73$ B.M. ($S = 1/2$).

In principle, the spin state and the magnetic moment can be easily determined by magnetic susceptibility measurements. In practice, however, a difficulty arises for a macromolecular compound such as haemoglobin. The diamagnetic correction for the ligand and the protein portion is substantial and difficult to evaluate. The conventional method to evaluate it is to measure the magnetic susceptibility of a diamagnetic derivative such as CO–Fe(II)–haemoglobin. However, a small correction for the temperature-independent paramagnetism in the diamagnetic Fe(II) derivative may be necessary. The electron paramagnetic resonance (EPR) technique is more powerful in determining the spin state and obtaining more detailed information about the electronic structure of paramagnetic compounds. EPR will be discussed in Experiment 20.

The electronic absorption spectra of porphyrin complexes are generally characterised by three peaks in the visible region; α, β and γ (Soret) bands. These bands are due to the π–π^* transitions in the porphyrin ring. The spectra of haem proteins are a little more complicated than the typical ones. Depending on the oxidation state and the spin state of the iron atoms, different charge transfer bands come into the spectra. A great amount of work has been done on the optical spectra of haem proteins; and the absorption spectra can now be utilised to identify and quantify the haem proteins.

MAGNETIC SUSCEPTIBILITY MEASUREMENTS

The principle of this technique has been detailed in Experiment 15. The only difference is that here we use a solution instead of powder as the sample. The volume susceptibility χ_v of a solution is approximated by

$$\chi_v = c_s\chi_s + (\rho - c_s)\chi_w \tag{18.8}$$

where χ_s and χ_w are the gram-susceptibility of the solute and the solvent, respectively, c_s is the weight concentration (g ml^{-1}) of the solute and ρ is the density of the solution. For water $\chi_w = -0.720 \times 10^{-6}$ c.g.s. The susceptibility of the solute is the one to be determined here. For a solution with a diamagnetic derivative such as carboxyhaemoglobin of the same concentration as above

$$\chi_v' = c_s\chi_s + (\rho - c_s)\chi_w \tag{18.9}$$

By subtracting equation 18.9 from 18.8, we get:

$$\chi_v - \chi_v' = (\chi_s - \chi_s')c_s = \chi_s^{corr} . c_s \tag{18.10}$$

χ_s^{corr} thus obtained can be regarded as the susceptibility value of the haem iron alone, being corrected for the diamagnetic contribution from the porphyrin ring and the globin protein.

χ_v and χ_v' are related to the weight changes, dW, in the following way:

$$\chi_v = 2\mathrm{d}W/aH^2, \ \chi_v' = 2\mathrm{d}W'/aH^2 \tag{18.11}$$

Therefore

$$\chi_v - \chi_v' = 2(\mathrm{d}W - \mathrm{d}W')/aH^2 \tag{18.12}$$

Using a standard solution for which χ_v is known, we can determine the necessary constant aH^2. We use water or an aqueous solution of $NiCl_2$ as the standard in this experiment.

Procedure

(i) Hang the designated empty tube (with cap) properly on to the magnetic balance. Measure its weight in, and out of, the magnetic field. Take several measurements and make sure that you are obtaining the proper values. Average the values. $dW_{tube} = W^{in}_{tube} - W^{out}_{tube}$.
(ii) Fill the tube to the mark with water, and make the measurements. $dW_{water} = W^{in}_{water} - W^{out}_{water}$.
(iii) Fill the clean and dry tube with the oxyhaemoglobin solution issued by a demonstrator and make the measurements. $dW_{oxy} = W^{in}_{oxy} - W^{out}_{oxy}$.
(iv) Add solid sodium dithionite to the oxyhaemoglobin solution in the sample tube (iii), and make measurements. $dW_{deoxy} = W^{in}_{deoxy} - W^{out}_{deoxy}$.
(v) Determination of the concentration of oxyhaemoglobin. Dilute the stock oxyhaemoglobin solution 250 times. Record the spectrum of the diluted solution. The concentration of the oxyhaemoglobin solution (g ml^{-1} = absorbance at 540 nm x 1.17 x 10^{-3} x dilution factor = absorbance at 560 nm x 1.89 x 10^{-3} x dilution factor (for bovine haemoglobin). If the solution is still too concentrated, dilute it further.
(vi) From a second batch of the stock solution, prepare methaemoglobin by adding a stoichiometric amount of potassium ferricyanide to the haemoglobin solution. Record all the necessary data such as the dilution factor owing to the addition of potassium ferricyanide solution. Fill the clean and dry tube to the mark with the methaemoglobin solution and make weighings. $dW_{met} = W^{in}_{met} - W^{out}_{met}$.
(vii) Add an excess of solid potassium cyanide to the methaemoglobin solution in the tube (vi), and make measurements. $dW^{in}_{CN} = W^{in}_{CN} - W^{out}_{CN}$.

From (i) and (ii), determine the apparatus constant. Assuming that the oxyhaemoglobin is diamagnetic and using dW_{oxy} as the diamagnetic correction for the haem and globin, calculate χ and μ values for deoxyhaemoglobin, methaemoglobin and cyanomethaemoglobin. In the calculation of (vi) and (vii), take the volume change due to the addition of potassium cyanide (if any) into consideration.

SPECTROPHOTOMETRIC STUDY OF AN EQUILIBRIUM CONSTANT

The equilibrium between aquo-methaemoglobin (H_2O-Fe(III)Hb) and hydroxomethaemoglobin (OH-Fe(III)Hb) will be studied, i.e.

$$\underset{(f)}{H_2O\text{-Fe(III)Hb}} \overset{K}{\rightleftharpoons} H^+ + \underset{(1-f)}{\text{OH-Fe(III)Hb}} \qquad (18.13)$$

The equilibrium constant K for equation 18.13 is determined here spectrophotometrically. K is expressed by

$$K = (H^+)(OH\text{-}Fe(III)Hb)/(H_2O\text{-}Fe(III)Hb) \tag{18.14}$$

Then,

$$\begin{aligned}\log K &= \log(H^+) + \log(OH\text{-}Fe(III)Hb)/(H_2O\text{-}Fe(III)Hb) \\ &= -pH + \log(1-f)/f\end{aligned} \tag{18.15}$$

Since the two forms show different absorption spectra, it is possible to determine $(1-f)/f$ from spectrophotometric data. The absorbance A_f at a wavelength λ of the solution containing the H_2O and the OH form of the total concentration c is given by

$$A_f = cf\alpha + c(1-f)\beta \tag{18.16}$$

where α and β are the molar extinction coefficients of the H_2O and the OH form, respectively. At limiting lower pH, $f = 1$, and, at high enough pH, $f = 0$. Therefore

$$A_1 = c\alpha, \; A_0 = c\beta \tag{18.17}$$

Hence, $(1-f)/f$ can be expressed by

$$(1-f)/f = (A_1 - A_f)/(A_f - A_0) \tag{18.18}$$

If one plots $(A_1 - A_f)/(A_f - A_0)$ against pH, one would obtain a straight line expressed by

$$\log(A_1 - A_f)/(A_f - A_0) = pH + \log K \tag{18.19}$$

Thus, one can determine $\log K$ from the intercept of graph 18.19.

Procedure

Into each of nine volumetric flasks, pipette 1 ml of the stock haemoglobin solution. Add about 90 ml of water and then add about 20 drops of a potassium ferricyanide solution (10 g in 500 ml) to each flask. Adjust the pH of the nine solutions to values ranging from 6.0 to 10.0 in increments of 0.5 ± 0.1 pH units according to the following method.

Pour one of the methaemoglobin solutions into a clean beaker (250 ml), and put the pH electrode in it. Depending on the pH desired, add either 100 mmol dm^{-3} HCl or NaOH by eye-droppers, while stirring continuously. Take care not to go much below pH 6 or above pH 10 during the adjustment process, since the protein may be denatured. Once the pH has been adjusted, the solution should be poured back into the volumetric flask whence it came and labelled with the pH reading. The volume then should be adjusted to the mark with water.

Record on a chart paper all the spectra of the nine solutions thus adjusted to pH. Select an appropriate wavelength (495, 560, 575 or 630 nm) and read off the necessary absorbance values. Plot your data according to equation 18.19 and determine K ($\log K$).

OXYGENATION CURVE BY SPECTROPHOTOMETRIC METHOD

The oxygenation curve, the fraction of oxyhaemoglobin as related to oxygen pressure, p, is determined spectrophotometrically for a haemoglobin solution. The absorbance A is related to the fraction y in the following way

$$A = cy\alpha_{oxy} + c(1-y)\alpha_{deoxy} \tag{18.20}$$

where c is the total concentration of haemoglobin and α_i is the molar extinction coefficient of species i. When $y = 1, A = A_{oxy} = c\alpha_{oxy}$ and when $y = 0$, $A = A_{deoxy} = c\alpha_{deoxy}$. Therefore

$$y = (A - A_{deoxy})/(A_{oxy} - A_{deoxy}) \tag{18.21}$$

Procedure

Obtain an oxyhaemoglobin solution (or appropriately diluted blood) prepared in pH 7.2 phosphate buffer. Add 10.0 ml of the solution to the tonometer as illustrated in *Figure VIII.18.2*. Connect the tonometer to an aspirator through a manometer, and evacuate it down to approximately 30 mm Hg; record the exact pressure reading. Close tap 2 and disconnect the manometer (and aspirator). Gently transfer the solution in flask 3 into the cuvette 4 and record the spectrum or read the absorbance at a certain wavelength, e.g. 730 nm. It may take as long as 25 min to get the equilibrium reading. A volume of air equal to the volume between the taps 1 and 2 is then introduced into the tonometer; to do this, first close tap 1 and then open tap 2. Swirl gently and equilibrate the solution and record the spectrum or read the absorbance at the set wavelength. Repeat this procedure 7–10 times.

Analysis of data

(i) Calculation of y

Ask the instructor for the value of $\alpha_{oxy}/\alpha_{deoxy}$ at the designated wavelength. Using this calculate y for each point.

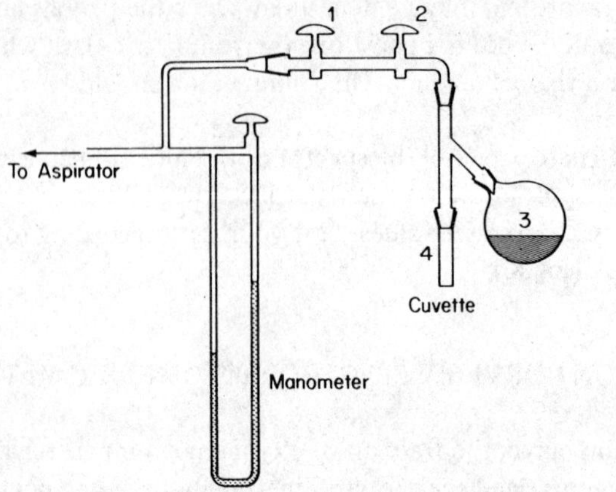

Figure VIII.18.2 Tonometer equipment; the volumes of the space between taps 1 and 2 and the dead volume of the space to the right of tap 2 must be known accurately

(ii) Calculation of p

The pressure of the oxygen in the initial experimental setup can easily be calculated from the fact that the reading on the manometer is $P = P_{air} + P_{water}$ and that the oxygen content of air is 20.4 vol%. The oxygen pressure in the subsequent experiment can be calculated knowing this initial oxygen pressure, the volume of the space between tap 1 and tap 2, and the dead volume of the rest of the tonometer.

(iii) Plot y against p using graph paper.

Is it hyperbolic or sigmoidal?

(iv) Scatchard method of estimating K_i

Define a parameter $Q = y/p(1 - y)$. Calculate Q for all the points and plot Q and ln Q against y. It has been shown that

$$\lim_{y \to 0} Q = K_1/4 \tag{18.22}$$

$$\lim_{y \to 1} Q = 4K_4 \tag{18.23}$$

$$\lim_{y \to 0} \mathrm{d}(\ln Q)/\mathrm{d}y = (8K_2 - 3K_1)/K_1 \tag{18.24}$$

$$\lim_{y \to 1} \mathrm{d}(\ln Q)/\mathrm{d}y = (8K_4 - 3K_3)/K_3 \tag{18.25}$$

The value of Q when extrapolated to $y = 0$ gives $K_1/4$ and when extrapolated to $y = 1$ gives $4K_4$. These two K_i values are thus fairly easily and accurately estimated. Using 18.24 and 18.25, one can also estimate K_2 and K_3 from the extrapolated values (to $y = 0$ and $y = 1$) of the tangent of the curve of $\ln Q$ versus y.

(v) Simulation of the experimental curve by the Adair equation

The curve-fitting may be made by assigning appropriate values for K_i with use of a computer or a programmable calculator. Since the Scatchard method gives a fairly good value for K_1, it is advisable to use this K_1 value for the simulation, and vary the other K_i values by trial and error.

REFERENCES

J. E. Falk (1964). *Porphyrins and Metalloporphyrins,* Elsevier, Amsterdam

W. S. Weissbluth *Structure and Bonding,* 2, 1 (1967); *Haemoglobin,* Springer-Verlag (1974)

D. W. Smith and R. J. P. Williams (1970). *Structure and Bonding,* 7, 1

Probes of Enzymes and Hemoproteins, Eds. B. Chance, T. Yonetani and A. S. Mildvan, Academic Press, New York (1971)

E. Antonini and M. Brunori (1971). *Haemoglobin and Myoglobin in Their Reactions with Ligands,* North-Holland Publishing Co., Amsterdam

E.-I. Ochiai (1977). *Bio-inorganic Chemistry, an Introduction,* Allyn and Bacon, Boston, Chapters 5 and 6

G. Buse (1971). *Angew. Chem., Int. Ed. Engl.,* **10**, 663

W. H. Sawyer (1972). *J. Chem. Educ.,* **49**, 777. (for Oxygenation Curve by Spectrophotometric Method)

Experiment 19
Oxygenation of Co(salen) – A Model for Haemoglobin

INTRODUCTION

Haemoglobin is vital to the human body in its function of combining with and carrying oxygen molecules. It consists of a protein called globin and a prosthetic group called haem. The latter is an Fe^{II} complex of protoporphyrin. The mechanism of the reversible oxygenation process of haemoglobin is, however, far from well understood, even after a history of study for almost a century (*see* the previous experiment for a review). Many simpler and so-called model compounds, which can act as O_2-carriers, have been developed in the hope of clarifying the mechanism; these include the complexes of Co^{II}, Fe^{II}, Ru^0, Os^0, Rh^I, Ir^I, Ni^0, Pd^0, and Pt^0. Co^{II} complexes are most numerous and perhaps the best understood, although how closely the mechanism of oxygenation of these is related to that of the haemoglobin system constitutes an intriguing and currently active research area.

Two types of O_2-addition for Co^{II} complexes have been distinguished:

$$Co^{II}L\ (L = \text{ligand}) + O_2 \overset{K_1}{\rightleftharpoons} O_2CoL \qquad \text{(A)}$$

$$2Co^{II}L'\ (L' = \text{ligand}) + O_2 \overset{K_2}{\rightleftharpoons} L'CoO_2CoL' \qquad \text{(B)}$$

That is, the compounds of type (A) absorb oxygen to a ratio $O_2:Co = 1.0$, while those of type (B) absorb to a ratio $O_2:Co = 0.5$. Both types of oxygenation can be reversed, releasing O_2 and regenerating the starting Co^{II} compound. This reverse deoxygenation reaction may be effected by heating the oxygenated complex (as solid or in solution), by bubbling an inert gas through a solution containing the O_2 complex or sometimes by adding a hydrocarbon or halogenated hydrocarbon solvent to the O_2 complex. Since reactions (A) and (B) are readily reversible, the rate and equilibrium position of the O_2-absorbtion reactions should depend on the O_2 concentration or the pressure of O_2, P_{O_2}.

For equilibrium (A)

$$K_1 = \frac{[CoO_2]}{[Co]\ [O_2]} = \frac{[CoO_2]}{[Co].kP_{O_2}} \qquad (19.1)$$

$$\text{If } n = \frac{[CoO_2]}{[Co]_{\text{total}}}$$

$$\frac{n}{1-n} = kK_1 . P_{O_2} \qquad (19.2)$$

Therefore, if $n/(1-n)$ is plotted against P_{O_2}, a straight line is obtained, the slope of which is kK_1.

For equilibrium (B)

$$K_2 = \frac{[CoO_2]}{[Co]^2\ [O_2]} = \frac{[CoO_2]}{[Co]^2.kP_{O_2}} \qquad (19.3)$$

and $$\frac{n}{(1-2n)^2} = kK_2P_{O_2}\ [Co]_{total} \qquad (19.4)$$

PREPARATION OF [Co(salen)]

Dissolve 2.34 g of salen in 120 ml of ethanol at 60–70°C in a 250-ml round bottom flask with a side-arm and stopcock attached. The side-arm is connected to an aspirator. Upon dissolution of the salen, quickly add 2.49 g of $Co(OAc)_2.4H_2O$ in 15 ml of water while briskly swirling the flask. Immediately stopper the flask, evacuate through the side-arm for a short time, then close the stopcock (be careful not to draw off the bulk of ethanol). Continue heating at 60–70°C with periodic swirling for 2 h. During this period, the initially formed brown solid slowly changes to brick red. Cool the solution to room temperature, collect the crystals (in air) on a sintered glass funnel, wash three times with 15 ml of water, and then dry in a vacuum desiccator over $CaCl_2$.

O_2-ABSORPTION BY [Co(salen)] IN DIMETHYLSULPHOXIDE (DMSO)

Procedure

This experiment determines the mole ratio $[O_2]_{absd}$: [Co(salen)] and its dependence on oxygen pressure (P_{O_2}). The apparatus used is shown in *Figure VIII.19.1*.

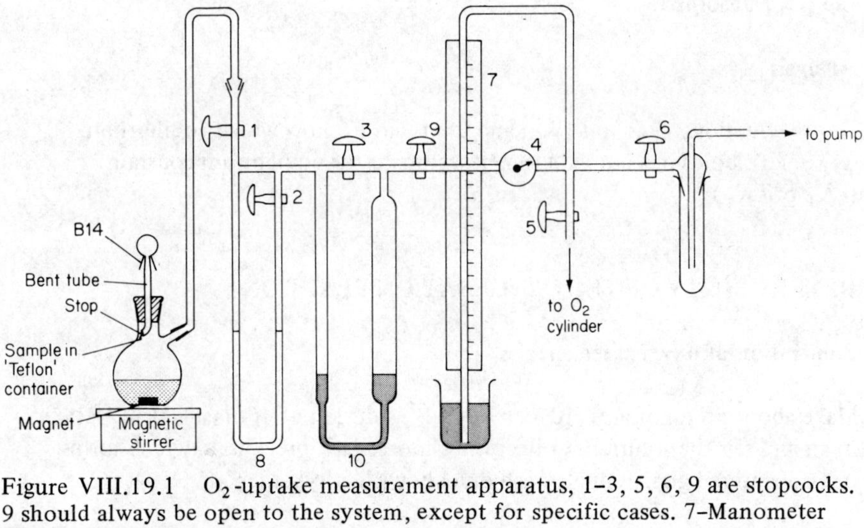

Figure VIII.19.1 O_2-uptake measurement apparatus, 1–3, 5, 6, 9 are stopcocks. 9 should always be open to the system, except for specific cases. 7–Manometer (Hg). 8–Manometer (oil). 10–Gas burette

Weigh accurately about 25-35 mg of [Co(salen)] into a small Teflon container which is slid through the top B14 joint of the reaction vessel containing 20 ml of DMSO and a stirrer magnet (make sure the bent tube is pointed at the stopcock 1, as in *Figure VIII.19.1*). Note that care must be taken in handling DMSO; although it is not itself poisonous, it is absorbed by the skin and can easily carry other compounds through the skin with it. Connect the flask to the measuring system and start vigorous agitation with the magnetic stirrer (allow about 0.5-in. space between the stirring motor and the vessel). Make sure that stopcocks 1-4 are open (5 closed) and then evacuate the system by opening 6. Flush a slow stream of O_2 through the two-way stopcock 5 to the atmosphere for a short time. Close 6 and then carefully open 5 to the system to introduce oxygen. Watch the mercury manometer 7 and stop the O_2 introduction at a certain O_2 pressure (about 170, 350, 520 and 700 mm Hg in four separate experiments). Allow a minute of stirring so that DMSO becomes saturated with O_2. Record the pressure (on manometer 7), and close 2-4. Introduce about 50 mm Hg more O_2 to the manometer side (7) of the apparatus through 5. Read the mercury height in the gas burette. You are now ready to start the gas uptake measurement. Drop the sample into the DMSO by turning the bent tube and start the stopclock; gas absorption will be indicated by the rising oil level in the left-hand limb of the oil manometer 8. Open 4 carefully and introduce further O_2 so as to make the oil levels of the manometer 8 even, and record the new height of the mercury level in the gas burette. Take this reading at time intervals of 30 s for 5-10 min. You can stop taking readings when there is no further absorption of oxygen. Plot the amount of O_2 absorbed versus time (record the reaction temperature, i.e. room temperature). Calculate the $[O_2]$ absorbed.

Analysis

Derive equations 19.2 and 19.4, and use them to show which equilibrium, (A) or (B), holds in your experiment; estimate the equilibrium constant (kK_1 or kK_2).

REVERSIBILITY OF THE OXYGENATION REACTION

Preparation of oxygenated species

Make about 50 ml of a *ca.* 10^{-2} mol dm^{-3} solution of [Co(salen)] in DMSO in air and stir the solution. Collect the almost black precipitate which forms upon a sintered glass funnel. Wash quickly with ethyl ether.

O_2 release

Try to dissolve the black precipitate into $CHCl_3$ and observe what happens.

Spectroscopic study

Make up 50 or 100 ml of the following solutions, record the visible spectra and compare them.

(i) About 10^{-4} mol dm^{-3} solution of [Co(salen)] in $CHCl_3$.
(ii) About 10^{-3} mol dm^{-3} solution of [Co(salen)] in DMSO, O_2 bubbled through the solutions for 5–10 min.
(iii) Dilute (ii) to about 1/20 with DMSO.
(iv) Dilute (ii) to about 1/20 with $CHCl_3$.

REFERENCES

E.-I. Ochiai (1973). *J. Inorg. Nucl. Chem.*, **35**, 1727–1739
C. F. Floriani and F. Calderazzo (1969). *J. Chem. Soc. A*, 946–953
R. H. Bailes and M. Calvin (1947). *J. Am. Chem. Soc.*, **69**, 1886–1893
R. G. Wilkins (1971). *'Bio-inorganic Chemistry'*, Eds. R. Dessy, J. Dillard and L. Taylor, *ACS Advances in Chem. Ser.*, **100**, 111–134
A. G. Sykes and J. A. Weil (1970). 'Inorganic Reaction Mechanisms', Ed. J. O. Edwards, *Progr. Inorg. Chem.*, **13**, 1–106
E. Antonini and M. Brunori (1971). *Haemoglobin and Myoglobin in their Reactions with Ligands,* North-Holland Publishing Co., Amsterdam

Experiment 20 EPR Study of Co(salen) and Co(salen)—O_2 Systems

INTRODUCTION TO EPR

EPR (electron paramagnetic resonance) or ESR (electron spin resonance) techniques are concerned with the resonance of the polarised spin or orbital magnetic moments of electrons. They can be fruitfully applied to systems containing unpaired electrons such as organic free radicals, triplet states of organic molecules, and compounds with metals having incomplete *d*-shells. Here discussion will be confined to *d*-electron systems. EPR may give detailed information about the oxidation state, the spin state, the ligand structure, and the energy levels of different electronic states of a *d*-electron system.

Suppose that an electronic spin system of S = ½ is brought under the influence of a static magnetic field H. The energy state of the system would split into two, $M_s = +½$ and $M_s = -½$, owing to the interaction of the spin system with the magnetic field (*see Figure VIII.20.1*). The energy gap between the two states is $\Delta E = E(-½) - E(+½) = 2\beta H$, where β is in Bohr magnetons, the unit of electron spin moment. The majority of the spins of the system are in the lower energy state $M_s = +½$. When an electromagnetic wave of frequency υ is applied to this system and a condition $\Delta E = h\nu$ is satisfied, an excitation from $M_s = +½$ to $M_s = -½$ can occur. As a result of this excitation, a fraction of the energy of the electromagnetic wave will be absorbed by this system. This is called 'electron paramagnetic resonance absorption'. For a free unpaired spin

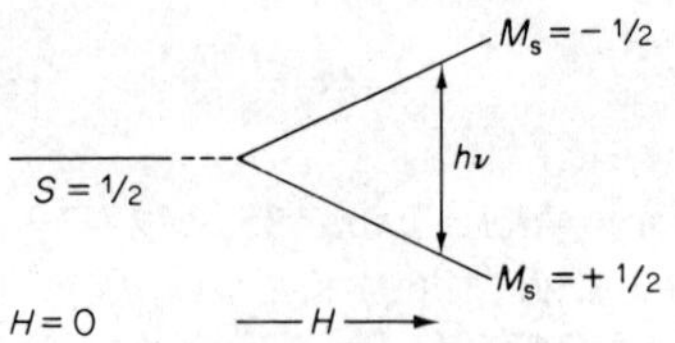

Figure VIII.20.1 Zeeman splitting and resonance absorption

$$\Delta E = h\nu = 2\beta H \tag{20.1}$$

Usually, a microwave of 3.2 cm wavelength (called the X-band) is used for EPR. Then $\nu \doteqdot 9200$ Mc s^{-1}. By putting explicit values for β and h, one gets $H = 3200$ gauss. The resonance condition is generally expressed by the following equation involving the g value:

$$h\nu = g\beta H \tag{20.2}$$

where

$$g = h\nu/\beta H = 7.1393/10^{10}\,\nu H \tag{20.3}$$

g is called the spectroscopic splitting factor or simply the g value; $g = 2$ (to be precise, 2.00229) for a single free electron. The g value should be 2.0 for any orbitally singlet ground state, because the orbital contribution to the magnetic moment in this case is zero. However, the g values tend to deviate from 2.0. This is because the ground state is in fact not purely orbitally singlet in many cases, owing to the admixture of excited states influencing the ground state through spin-orbit coupling. Values of g for a formal orbitally singlet ground state can be expressed by:

$$g = 2(1 - n\lambda/\Delta) \tag{20.4}$$

λ is the effective spin-orbit coupling constant, Δ is the energy gap between the ground level and the excited level in question, and n is a certain constant depending on the two levels concerned. Thus, the g value experimentally determined could afford information about the effective value of λ and the energy gap Δ. The former, λ, may be interpreted to give a measure of the covalency of the bond between the central metal atom and the ligands. The g value can vary from one direction to another; i.e. if x-, y-, and z-axis are defined in a complex, g_x, g_y and g_z may not be the same. They are the same for a strictly octahedral complex; $g_x = g_y = g_z$, i.e. isotropic. If the structure of the complex is tetragonal: $g_x = g_y (= g_{\perp}) \neq g_z (= g_{\parallel})$. If the symmetry of the complex is lower than axial, $g_x \neq g_y \neq g_z$. Therefore, the fundamental symmetry of the ligand field may be deduced from the shape of the EPR spectrum.

The electron spin, which is a magnetic moment, can also interact with nuclear spins of the central metal atom on which the spin is mainly located, and the nuclear spins of the ligand atoms. These interactions give the EPR spectrum a hyperfine structure (owing to the metal atom) or a super hyperfine structure (owing to the ligand atoms). Because of the selection rule regarding the hyperfine transitions, the hyperfine structure consists of (almost) equally spaced $(2I + 1)$ lines; here I is the quantum number of the nuclear spin. The intervals between the hyperfine lines, A, are called hyperfine parameters, and are related to the effective electron density around the nucleus and other factors. The A value is also variant with the direction; e.g. in a tetragonal symmetry $A_x = A_y (= A_{\perp}) = A_z (= A_{\parallel})$.

The EPR spectrum can be obtained with samples in solution, in the frozen state and in the single-crystal state. The samples in solution give time-averaged spectra, because of the rapid molecular motion. The single-crystal EPR spectra can yield the most detailed information, but practically it is the most difficult to take. The frozen samples are made by cooling and freezing suitable sample solutions. The molecules in question are assumed to be completely randomly orientated. The EPR absorption spectrum of a frozen sample of species of tetragonal symmetry ($g_{\perp} \neq g_{\parallel}$) has been shown to have such a shape as illustrated in *Figure VIII.20.2(A)*. In the case of EPR, the first derivative, (b, *Figure VIII.20.2(A)*), rather than the original absorption, (a, *Figure VIII.20.2(B)*), is recorded on the instrument. The g values may be

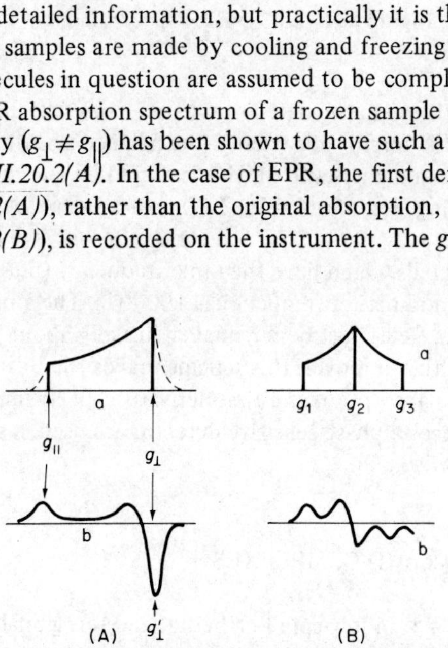

Figure VIII.20.2 Frozen (powder) spectrum

determined on the frozen (or powder) spectra as illustrated. However, there remains an uncertainty in assigning the $g_\perp$. Empirically speaking, $g'_\perp$ as shown in *Figure VIII.20.2(A),* gives a better estimate for rather sharp spectra, but the true value is closer to $g_\perp$ in rather broad spectra. To be precise, the spectra must be computer-simulated.

INSTRUMENTATION

The basics of the EPR spectrometer is illustrated in *Figure VIII.20.3.* The microwave is emitted from a klystron. The magic tee is theoretically adjusted such that no microwave energy comes out to the crystal detector when no absorption occurs in the sample cavity. However, in practice, the magic tee is so adjusted

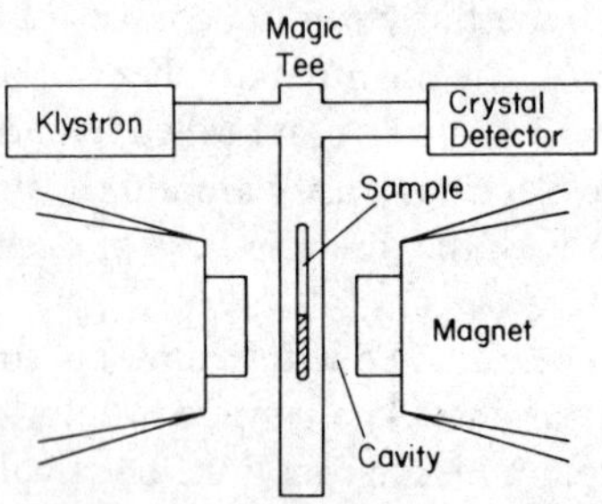

Figure VIII.20.3 EPR spectrometer

that a small current flows in the detector when no absorption takes place. When the sample absorbs the microwave energy, the balance is shifted and as a result the detector picks up a higher current; this is amplified and recorded. The magnetic field is changed continuously. To increase signal-to-noise ratio, the magnetic field is normally modulated and a phase-sensitive detector is used to detect only those signals which have the same frequency (phase) as a reference. In many cases, the modulation frequency is 100 kHz. The amplitude of the modulating magnetic field must be far smaller (usually about 1/10) than the absorption peak width. Moreover, this scheme makes the EPR spectrometer record the resonance absorption as a first-derivative spectrum. If a second modulation and a second phase-sensitive detector are used, a second-derivative curve can be obtained.

EPR OF LOW-SPIN Co(II) COMPLEXES

Co(II) has d^7 and $S = \frac{1}{2}$ in low-spin octahedral and tetragonal complexes. In a rhombic and tetragonal field, the probable electronic structure is $(d_{yz})^2(d_{zx})^2(d_{x^2-y^2})^2(d_{z^2})^1(d_{xy})^0$; the x- and y-axis are defined so that they bisect the angles ligand–metal–ligand. Theoretically, the g values and the hyperfine parameters A for this electronic configuration are given by the following equations

$$\begin{cases} g_x = 2 - 6a_1 \\ g_y = 2 - 6a_2 - 3a_1^2 \\ g_x = 2 - 3a_1^2 \end{cases} \tag{20.5}$$

$$\begin{cases} A_x = P(-6a_1 - 2/7 - 3a_2/7 - k) - 3Pa_1^2/14 \\ A_y = P(-6a_2 - 2/7 - 3a_1/7 - k) \\ A_z = P(+4/7 + 3(a_1 + a_2)/7 - k) - 9Pa_1^2/14 \end{cases} \tag{20.6}$$

where $a_1 = \lambda/(E(d_{z^2}) - E(d_{yz}))$, $a_2 = \lambda/(E(d_{z^2}) - E(d_{zx}))$, $P = 2\beta\beta_N < r^{-3} >$. When a_1^2 is negligibly small, the terms including a_1^2 can be omitted. If the ligand field is strictly tetragonal

$$g_x = g_y = g_\perp = 2 - 6a;\ g_z = g_\parallel = 2.0 \tag{20.7}$$

In the case of Co(II), λ is negative and hence $g_x > g_y > g_z$. The EPR spectra can thus be illustrated as shown in *Figure VIII.20.4.* To complicate the matter, the hyperfine interaction with the nucleus of the Co atom, which has $I = 7/2$, splits each band into eight peaks. In most cases, A_x and A_y ($A_\perp$) are small and the hyperfine structures on g_x and g_y ($g_\perp$) are not well-defined; while the hyperfine on g_z ($g_\parallel$) is well-separated peaks (*Figure VIII.20.4*).

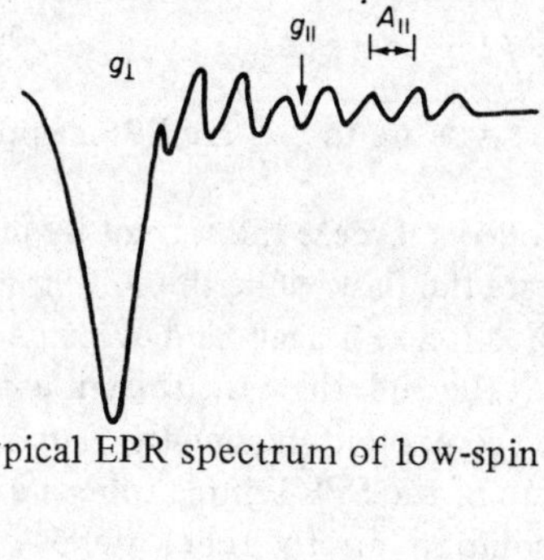

Figure VIII.20.4 Typical EPR spectrum of low-spin Co(II) complex of tetragonal symmetry

A rough estimate of g_i and A_i can be obtained from the spectrum as illustrated in *Figure VIII.20.4*. To be exact, a computer-simulation must be performed. From these values, a_i, k and P values may be estimated, using equations 20.6 and 20.7. The ratio λ_{obs}/λ_0 (λ_0 for the ion *in vacuo*) may be approximated to P_{obs}/P_0 (P_0 for the ion *in vacuo*). λ_{obs} and P_{obs} give an estimate of the extent of the delocalisation of the unpaired electron. If a reasonable estimate of λ_{obs} is obtained from the above approximation, the energy gaps

($E(d_{z^2})-E(d_{yz})$, etc.) may be evaluated. When an electronic absorption spectrum of the complex is available, it may allow an estimation of those energy gaps. Then the data on the electronic structure obtainable from the EPR may be compared to those obtained from the electronic spectrum.

In this experiment, the EPR spectrum of the so-called (1:1) O_2-Co complex will also be studied. The electronic structure of the oxygenated species can be approximated by O_2^--Co(III). The O_2^- radical should have an axial symmetry in the free state, and it could become rhombic when bound to Co(III).

PROCEDURE

Preparation of sample solution

Add about 10 ml of toluene-pyridine (90–10) mixture to a specially designed 50-ml flask *Figure VIII.20.5.* Hang a bucket containing about 3 mg of Co(salen), and attach the EPR sample tube evacuated, as illustrated.

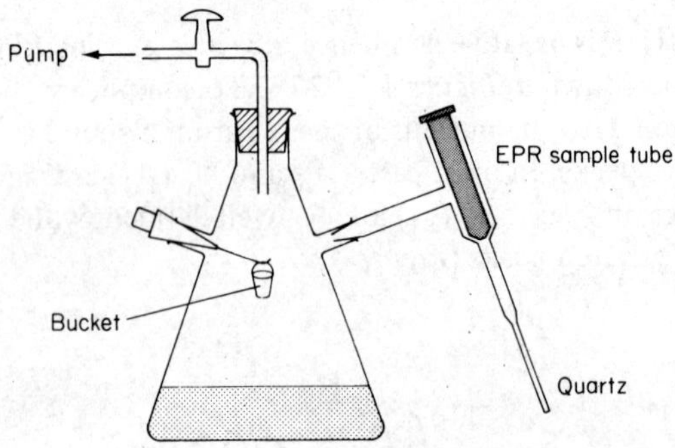

Figure VIII.20.5 A setup to prepare EPR sample under vacuum

De-gas the solvent as follows. Freeze the solvent by dipping the flask into liquid nitrogen and evacuate the flask while the solvent is solid. Thaw the solvent completely and then freeze it again and evacuate the system. Repeat this at least three times. At the end, thaw the solvent completely, lower the bucket and dissolve all the compound. Incline the flask so that the solution comes to the top of the tap of the EPR sample tube and open the tap slowly to introduce the sample solution into the tube. Add 5–7 cm length of solution and close the tap and disconnect the sample tube from the flask.

Taking EPR spectra

The detailed operational procedure for the specific instrument to be used will be posted beside the machine.

Place the Dewar flask filled with liquid nitrogen in the cavity and put the sample tube into it. If necessary, blow nitrogen or other gas through the cavity, so that moisture may not gather inside. Set the switch to tuning position and adjust the klystron frequency to the centre of the dip on the oscillograph. Switch to measuring position and adjust the attenuator (to, for example, 5 mV) as well as the magic tee to bring the needle on the detector meter in the designated region. Set the field centre to, for example, 3250 G (for $g = 2.0$) and the scanning width to, for example, 2×10^3 G at the beginning, and modulation width to, for example, 10 G. These should be optimised by trial and error. Adjust the recorder gain control so that the spectrum spans the width of the chart paper and record the spectrum. For calibration, a DPPH ($g = 2.0036$) sample may be added externally to the cavity and its peak position may be recorded on the same chart.

After recording the spectrum under proper conditions noting all necessary data, bring the sample tube to room temperature. Open the tap for a second or so, close it, and shake the tube by inverting it. Immediately put it back in liquid nitrogen in the Dewar flask. Co(salen)py_n has now been oxygenated, mostly to the O_2-Co form. Record the EPR spectrum of the sample.

ANALYSIS OF SPECTRA

Read off the magnetic field H_i from the spectra and make corrections, if necessary. Calculate g_i values according to equation 20.3. Convert $|A_i|$, in units of gauss, to $|A_i|$, in units of $\times 10^{-4}$ cm^{-1}. Applying equations 20.5-20.7, estimate a_i, P and k as far as possible ($P_0 = 220 \times 10^{-4}$ cm^{-1}). Discuss the symmetries of the complexes, and their electronic structures as far as possible.

REFERENCES

G. E. Pake and T. L. Estle (1972). *The Physical Principles of Electron Paramagnetic Resonance,* Benjamin, 2nd edn.

P. B. Ayscough (1967). *Electron Spin Resonance in Chemistry,* Methuen

R. G. Wilkins (1971). *Adv. Chem. Ser.,* **100,** 111

B. M. Hoffman, D. L. Diemente and F. Basolo (1970). *J. Am. Chem. Soc.,* **92,** 61

E.-I. Ochiai, *J. Inorg. Nucl. Chem.,* **35,** 1727, 3375 (1973); **37,** 351 (1975)

Experiment 21
Vitamin B_{12} and Model Compounds

INTRODUCTION

Vitamin B_{12} is fairly widely distributed in biological systems; lack of it is known to cause pernicious anaemia and other symptoms. It is a very interesting compound from the viewpoint of its inorganic chemistry.

Vitamin B_{12} (VB_{12}), which is also called cyanocobalamin, is a dark-red coloured cobalt(III) complex containing a large macrocyclic corrin ligand. The corrin ring has four coordinating nitrogen atoms which form roughly a square plane around the central cobalt atom. One of the axial positions (fifth position) is occupied by 5,6-dimethylbenzimidazole, and the other (sixth position) by a cyanide group in the case of cyanocobalamin. Many derivatives of VB_{12} can be obtained by replacing the sixth axial ligand. If CN^- is replaced by H_2O, the resulting compound is aquocobalamin or VB_{12a}; if 5′-deoxy-adenosyl group occupies the sixth position, the derivative is called VB_{12} co-enzyme. This co-enzyme is, in fact, the most important derivative with respect to the biological function of VB_{12}. The structure of VB_{12} co-enzyme and some other derivatives of VB_{12} are shown in *Figure VIII.21.1*. VB_{12}

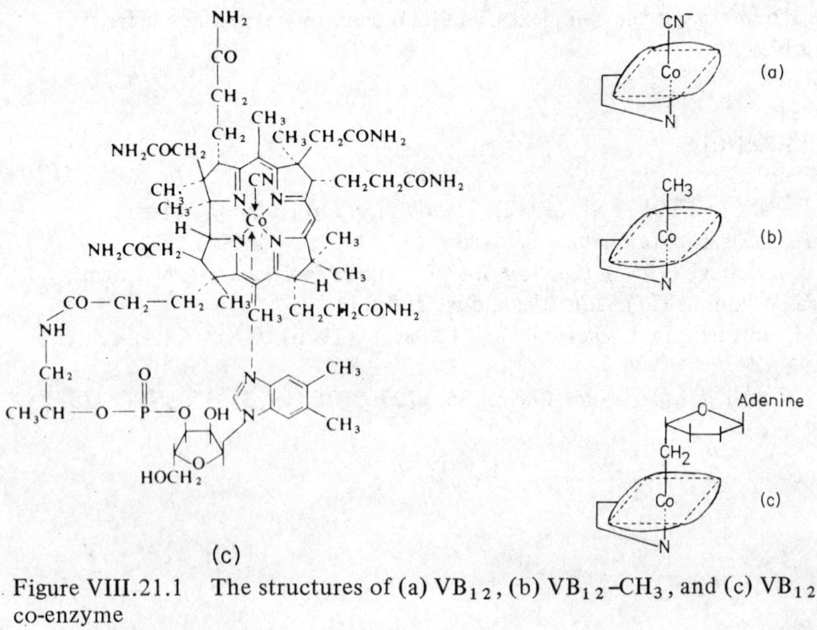

Figure VIII.21.1 The structures of (a) VB_{12}, (b) VB_{12}–CH_3, and (c) VB_{12} co-enzyme

derivatives are known to contain cobalt in different oxidation states, I, II, III, and the systems are highly susceptible to redox processes at the metal centre.

VB_{12} may be reduced by a number of reducing agents to brown VB_{12r} (Co(II)), and then to grey-green VB_{12s}(Co(I)). VB_{12s} reacts with a variety of alkyl halides and their analogues to give alkyl derivatives of VB_{12}. Some of the

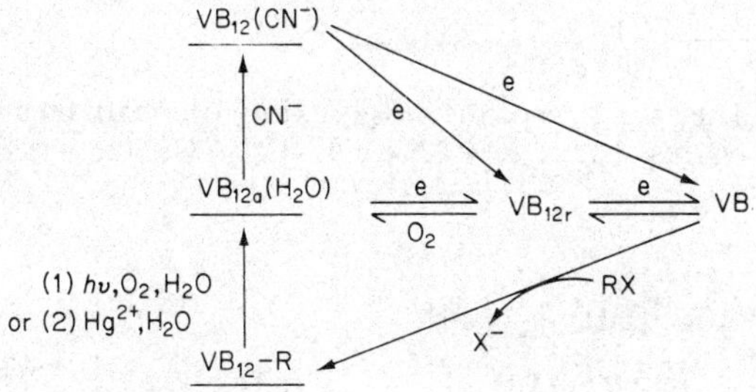

Figure VIII.21.2 Reactions of VB_{12}

more important reactions are summarised in *Figure VIII.21.2*.

Some of the important reactions which VB_{12} and its derivatives participate in are:

$HOOCH_2CH_2CH(NH_2)COOH \rightleftharpoons HOOCCH(CH_3)CH(NH_2)COOH$ (21.1)
glutamic acid — β-methylaspartic acid

$CH_3CH(OH)CH_2(OH) \rightarrow CH_3CH_2CHO + H_2O$ (21.2)
1,2-propanediol

$HSCH_2CH(NH_2)COOH \rightarrow CH_3SCH_2CH(NH_2)COOH$ (21.3)
homocystein — methionine

$VB_{12}\text{-}CH_3 + HgX^+ \rightarrow VB_{12}(H_2O) + XHgCH_3$ (21.4)

The last reaction (21.4) has been discovered very recently, and plays a very important role in converting inorganic mercury compounds to organomercury compounds. Methyl mercury compounds ($(CH_3)_2Hg$, $(CH_3)HgX$) have been revealed to be highly toxic, particularly in brain tissue. The possible role of a reaction such as 21.4 in the mechanism of mercury pollution remains to be fully established.

Since VB_{12} is highly complicated, several model compounds for it have been devised. One such series is the Co-dimethylglyoxime complexes (cobaloximes). Here we prepare a few derivatives of cobaloxime (shown in *Figure VIII.21.3*), and try a few reactions which are similar to those mentioned above. Some of the reactions in these preparative reactions are outlined below (where DH_2 = dimethylglyoxime).

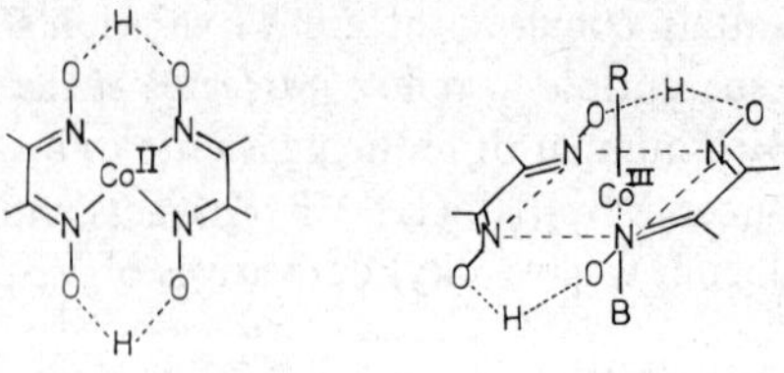

(a) (b)

Figurė VIII.21.3 The structure of (a) bis(glyoximato)cobalt(II) and (b) cobaloxime derivatives where R = CH_3, CH_2CH_2CN, etc., and B = pyridine, H_2O, etc.

$$\left\{\begin{array}{l} Co^{II} + 2DH_2 \rightarrow \tfrac{1}{2}[Co^{II}(DH)_2]_2 + 2H^+ \\ [Co^{II}(DH)_2]_2 + H_2 \rightarrow 2[Co^{I}(DH)_2]^- + 2H^+ (OH \rightarrow H_2O) \\ [Co^{II}(DH)_2]_2 \rightleftharpoons [Co^{I}(DH)_2]^- + [Co^{III}(DH)_2] \\ [Co^{I}(DH)_2]^- + CH_2{=}CHCN \rightleftharpoons \left(\begin{array}{c} CH_2{=}CHCN \\ \vdots \\ Co^{I}(DH)_2 \end{array}\right) \underset{OH^-}{\overset{+H^+}{\rightleftharpoons}} \left\{\begin{array}{l} (DH)_2Co{-}CH_2CH_2CN \\ \quad\text{or} \\ (DH)_2Co{-}\underset{\displaystyle CH_3}{CH}CN \end{array}\right. \end{array}\right. \qquad (21.5)$$

(A) STUDY ON VITAMIN B_{12}

Polarographic study of VB_{12}

Principles of polarography instrument

Polarography and voltametry are analytical methods where the basic process is a simple electrolysis. In the case of polarography, this occurs at a dropping mercury electrode where mercury drops through a capillary from a mercury reservoir into the electrolyte solution. With a three-electrode system, the circuit is completed by a counter electrode (a Pt wire) and the emf applied to the dropping mercury electrode (DME) is measured versus a separate calomel electrode. In principle, the instrument applies a varying emf to the DME and measures (and plots) the current flow caused by the reduction or oxidation of various species present in the solution. In oversimplified terms, the circuitry is shown below in *Figure VIII.21.4*, although in modern instruments the emfs are varied electronically rather than mechanically.

Variable resistance R_1 regulates the total potential applied from the DC source to rheostat R_2. R_1 controls the range of emf values to be scanned by R_2, from which the emf is applied across the DME and the counter electrode. An accurately known resistor, R_S, is in the circuit leading to the counter electrode

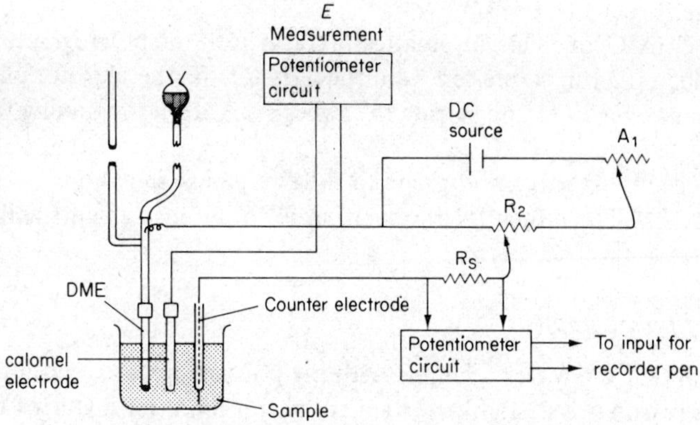

Figure VIII.21.4 Polarographic circuit

and the potential drop E across R_s is measured. Since $i = E/R_s$, this potential gives a value for the current flowing to the cell and this can be applied to a recorder pen. The emf applied to the DME is measured with a potentiometer inserted between the DME and a calomel electrode.

General procedure for operating the instrument

Attach the cell to the dropping mercury electrode (DME) assembly. Adjust the mercury reservoir height for a drop rate of about 5 s. The following is a general operating procedure applicable to all runs.

(1) Turn on both the AC and DC switches. At least 30 min should be allowed for warm-up before proceeding to step (2).

(2) Balance amp (1) by placing the function switch on 'BAL 1' and adjusting the meter needle to 0 with the amp (1) BALANCE control.

(3) Similarly balance amps (2), (3) and (4) using switch positions BAL 2, 3 and 4 and amp (2), (3) and (4) BALANCE controls, respectively. Check these before each run. Then set the function to CELL.

(4) Set the INITIAL POTENTIAL dial (0.00 V for the first run).

(5) Set the SWEEP RATE (0.2 V min^{-1} for the first run). SWEEP DIRECTION is '–' for all runs.

(6) Set the CURRENT sensitivity (5 μA full scale for the first run).

(7) CURRENT COMPENSATION knobs are off for all runs.

(8) SWEEP switch is left ON permanently.

(9) Set DAMPING ('3' for the first run).

(10) Turn ON chart recorder POWER and SERVO. CHART is OFF.

(11) Recorder SENSITIVITY is on 100 mV and the left knob is set to CALIBRATED.

(12) Adjust the pen to the *LEFT* of the chart with the ZERO knob (ignore the chart calibrations which do not apply when this general purpose recorder is used in this system).

(13) Set chart speed to 1 $\frac{\text{INCH}}{\text{MIN}}$.
(14) Turn CHART on. This automatically starts both the polarograph and the recorder. When the run is finished, turn the CHART off thereby stopping both polarograph and recorder and setting the sweep back to the beginning for the next run.

(This procedure is only appropriate to the equipment shown in *Figure VIII.21.4.* The principles of operation for other instruments will differ somewhat.)

Principles of polarography

It is essential that the waves obtained with the polarograph are reproducible, and for this reason it is desirable to standardise the conditions. One of the factors which influences the flow of current through the cell is the concentration of the electroactive component. When an ion reaches the dropping electrode under the conditions of diffusion control, the height of the polarographic wave is exactly proportional to the amount of the electroactive ion. To ensure these conditions, a supporting electrolyte is added to the solution so that most of the current (up to 90%) will be transported by the ions of the supporting electrolyte. The supporting electrolyte must be composed of ions which are discharged at far higher potentials (more negative potentials) than the ions under investigation, and which will not interfere with them.

When an excess of supporting electrolyte is present in the solution, the electrical force on the reducible ions is nullified. Under this condition, the limiting current is almost a *diffusion current* (i_d). Ilkovič examined the various factors which govern the diffusion current, and deduced the following equation

$$i_d = 607nD^{1/2}C\,m^{2/3}\,t^{1/6} \qquad (21.6)$$

where i_d = diffusion current in mA;
n = moles of electrons involved in the electrochemical reaction of one mole of the electroactive species;
D = diffusion coefficient of the species in $cm^2\ s^{-1}$;
C = concentration of the species in mmol dm^{-3};
m = rate of flow of mercury from the dropping electrode in mg s^{-1};
t = drop time in s.

Another important term, half-wave potential $E_{1/2}$, must be defined. This is simply the potential corresponding to a current that is equal to one-half the diffusion current. In polarography, this potential is taken in the middle of the polarographic wave (*see Figure VIII.21.5*). The half-wave potential is a characteristic constant for a specific redox system.

Consider the following reduction process on a dropping electrode (cathode):

$$Ox^{n+} + ne \rightleftharpoons Red \qquad (21.7)$$

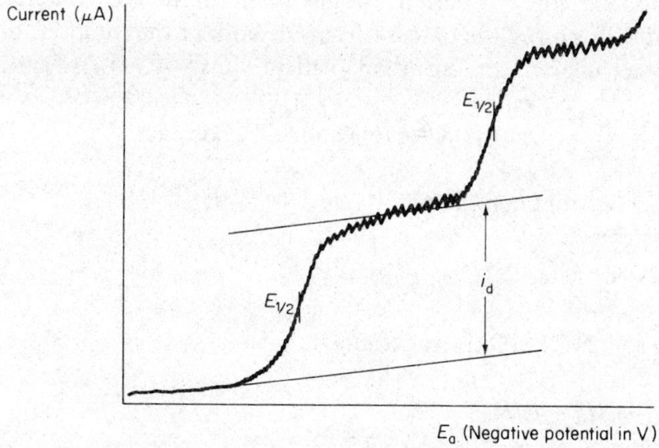

Figure VIII.21.5 A polarogram

For such a process, if reversible, the electrode potential is given by the Nernst equation 21.3:

$$E = E^0 + \frac{RT}{nF} \ln \frac{A_{ox}}{A_{red}} \tag{21.8}$$

where A_{ox} and A_{red} = activities of the oxidant and reductant;
R = gas constant;
T = absolute temperature;
n = number of electrons;
F = Faraday constant = 965 000 coulomb;
E^0 = standard redox potential.

In a polarographic study, the activities may be substituted by concentration terms without appreciable error. The concentration here must be the one at the electrode surface, denoted by subscript '0'. Thus

$$E = E^0 + \frac{RT}{nF} \ln \frac{[\mathrm{Ox}]_0}{[\mathrm{Red}]_0} \tag{21.9}$$

Before commencement of the polarographic wave, only a small residual current flows, and the concentration of any electroactive substance must be the same at the electrode interface as in the bulk of the solution. As soon as the decomposition (electrolysis) potential is exceeded, some of the reducible species at the interface is reduced, and must be replenished from the body of the solution by means of diffusion. The reduction product does not accumulate at the interface, but diffuses away from it into solution or into the electrode material. When the applied potential is increased to a value at which all the oxidant reaching the electrode surface is reduced, then the current flowing will be the diffusion current. The current i at any point on the wave is determined by the rate of diffusion of the oxidant from the bulk of the solution to the electrode surface under a concentration gradient ($[Ox] - [Ox]_0$). That is:

$$i = K([Ox] - [Ox]_0),\ K = \text{proportionality constant} \tag{21.10}$$

Under diffusion current conditions

$$i = K[Ox] = i_d \tag{21.11}$$

From 21.10 and 21.11, $[Ox]_0$ is obtained:

$$[Ox]_0 = (i_d - i)/K \tag{21.12}$$

The reductant, if not originally present with the oxidant, will diffuse from the surface of the electrode to the bulk of the solution. Hence

$$i = k[Red]_0 \text{ or } [Red]_0 = \frac{i}{k} \tag{21.13}$$

Substitution of $[Ox]_0$ and $[Red]_0$ in 21.9 by 21.12 and 21.13 gives

$$E = E^0 + \frac{RT}{nF} \ln \frac{k}{K} + \frac{RT}{nF} \ln \frac{i_d - i}{i} \tag{21.14}$$

$$= E^{0\prime} + \frac{RT}{nF} \ln \frac{i_d - i}{i} \text{ (where } E^{0\prime} = E^0 + \frac{RT}{nT} \ln \frac{k}{K}) \tag{21.15}$$

When $i = i_{d/2}$

$$E = E_{1/2} = E^{0\prime} = \text{constant} \tag{21.16}$$

At 25°C

$$E = E^{0\prime} + \frac{0.059}{n} \log \frac{i_d - i}{i} \tag{21.17}$$

Polarography of VB$_{12}$

(a) Arrange the capillary so that the tip is immersed in 100 mmol dm^{-3} K_2SO_4 solution contained in a small beaker. Collect 30 drops of mercury as they fall from the capillary. Wash the mercury with water then acetone, dry in a stream of cold air, transfer to a previously weighed weighing bottle and weigh. Set up the complete cell (*see Figure VIII.21.4*) with 100 mmol dm^{-3} K_2SO_4 solution as electrolyte, and measure the time for 10 drops of mercury to fall from the capillary at an applied voltage of -1.4 V versus the calomel electrode. Take the average of three readings. Hence calculate the capillary characteristics m and t in the Ilkovič equation 21.6.

(b) Take a polarogram of 100 mmol dm^{-3} K_2SO_4 solution by adding 20 ml of 100 mmol dm^{-3} K_2SO_4 solution to the polarographic cell and then add 0.5 ml of Triton solution. Set up the complete cell and run a polarogram in the range 0–2.5 V.

The polarogram obtained shows two waves corresponding to the two reduction steps of oxygen:

a. $\left.\begin{array}{l} O_2 + 2e \rightleftharpoons O_2^- \\ O_2 + 2H^+ \rightleftharpoons H_2O_2 \end{array}\right\}$ 1st wave

b. $H_2O_2 + 2e \rightleftharpoons 2OH^-$ 2nd wave

The solution will now be degassed. Bubble N_2 via a sinter through the electrolyte for at least 10 min. Make another trace using the same settings as before.

(c) Take a polarogram of VB_{12} by preparing a 1 mmol dm^{-3} solution (25 ml) of VB_{12} in 100 mmol dm^{-3} K_2SO_4. Take the polarogram using 20 ml of the solution (with 0.5 ml of Triton X) after degassing, with the appropriate settings. Plot against log $(i_d - i)/i$ and hence find $E_{½}$ and n. Use the Ilkovič equation (21.6) to calculate the n value from the diffusion current. The diffusion coefficient (D) of VB_{12} under the experimental conditions is 2.95×10^{-6} cm^2 s^{-1}.

Spectroscopic study of the reaction of VB_{12}

Take a few ml of the VB_{12} solution prepared for the polarographic study, and dilute about 20 times. Record the spectrum in the range 800–320 mm. Now close the cell with a special stopper provided, and degas with nitrogen (more than 20 min), using the needles provided as inlet and outlet for the gas. Quickly open and add 1 mg of potassium borohydride, and then stopper and continue passing nitrogen until the solution becomes greenish (very pale when dilute). Record the spectrum of this solution (VB_{12s}) on the same chart. The base line may be higher than expected, owing to decomposition of the excess borohydride, which gives rise to bubbling. Try to eliminate the bubbles as much as possible.

Inject one drop of methyl iodide into this solution by a syringe, taking care to shield the cell from the light. Record the spectrum. Now expose the red methyl cobalamin solution in the cell to air and light. After 1–2 h take the spectrum. The final product should be VB_{12a} (*see Figure VIII.21.1*).

(B) METHYLATION OF MERCURY BY METHYL COBALAMIN

The reaction between Hg(II) and methyl cobalamin is considered to proceed according to the following scheme.

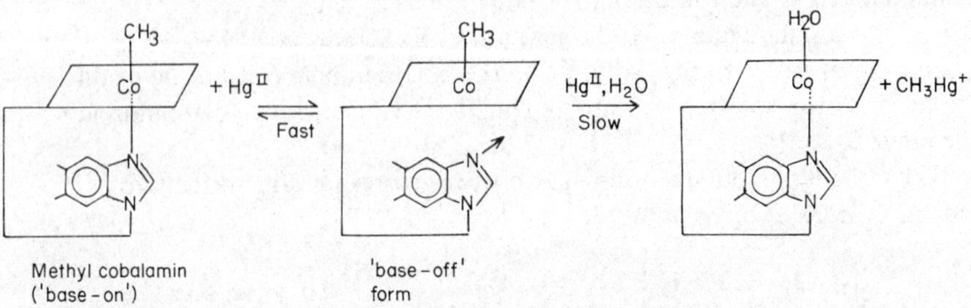

The 'base-off' form shows an absorption spectrum different from that of the 'base-on' form, having an absorption maximum at 480 nm instead of the 520 nm of the 'base-on' form. In this experiment, the reaction is followed spectroscopically.

Procedure

A solution of cyanocobalamin (100 mg) and cobalt nitrate (1 mg) in water (10 ml) is placed in a 25-ml three-neck flask. The solution is deoxygenated for 10 min by bubbling with hydrogen. After 10 min, a deoxygenated solution of sodium borohydride (20 mg) in water (0.5 ml) is added. Bubble hydrogen for an additional 5 min and then add 200 mg of methyl iodide. The cobalamin solution is purified by extraction through phenol (*see* the procedure following), and the resulting aqueous solution is reduced to 1-2 ml on a flash (rotary) evaporator (bath temperature $< 50^\circ C$). Add acetone dropwise until the solution shows a faint turbidity. Leave the solution overnight, preferably in a refrigerator, for the product to crystallise.

Collect the crystals by filtration, wash with acetone and determine the weight of the product. Make up a solution (approximately 2×10^{-4} mol dm^{-3}) of methyl cobalamin. (Note this solution is light-sensitive and must be handled in the dark or in a flask wrapped in black paper.) Pipette 2 ml of this solution into a UV-visible cell and add 0.05-0.1 ml of 3 mmol dm^{-3} $Hg(OAc)_2$. Start the timer at the time the mercury solution is added and record a spectrum immediately and then record the spectra every 150 s. Record the spectra, one on top of the other, on the same piece of paper. The time interval may be increased to 5 min after the reaction has slowed down.

Extraction through phenol

Prepare a stock solution of phenol (50 g) in methylene chloride (50 ml). Extract the aqueous cobalamin solution with 2 ml of the phenol stock solution. Separate the organic layer and re-extract the aqueous layer with successive aliquots of phenol–methylene chloride solution until no further colour is extracted. The combined organic extracts are washed with distilled water (twice with one-fifth volume of the organic layer) and the organic layer is diluted with methylene chloride to 10 times its original volume. The cobalamin is re-extracted from the organic layer with aliquots of distilled water (one-twentieth the volume of the organic layer) until no colour remains in the organic layer. The combined aqueous extracts are then washed with methylene chloride (three times the volume of the combined aqueous layer) to remove traces of phenol.

(C) STUDY ON MODEL COMPOUNDS

Preparation of α-cyanoethylpyridinatocobaloxime(III)

Add 2.32 g of dimethylglyoxime to 60 ml of methanol in a 125-ml Erlenmeyer flask with a side-arm for gas inlet. In a fume hood, bubble H_2 gas through the solution and add 2.49 g of $Co(OAc)_2.4H_2O$ and then 1.5 ml of acrylonitrile. H_2-bubbling is continued at room temperature for 1.5 h. Filter off the unreacted material (in air) and add 100 ml of water to the filtrate and then 1 ml of pyridine. Remove methanol using a rotary evaporator (do not heat above 40°C). Yellow crystals will separate. Collect the crystals on a sintered glass funnel, wash three times with 10 ml of water and air-dry. This compound may be analysed for C, H and N. Record its IR spectrum using Nujol mull, and its NMR spectrum (in $CHCl_3$ spectroscopic grade).

Preparation of β-cyanoethylpyridinatocobaloxime (III)

Add 2.38 g of $CoCl_2.6H_2O$ and 2.32 g of dimethylglyoxime to 60 ml of methanol in a 125-ml Erlenmeyer flask with H_2 gas bubbling through, and cooled in an ice bath. After complete dissolution of $CoCl_2$ (it may be necessary to momentarily warm the flask to room temperature), add 1.5 ml of acrylonitrile and then a mixture of 1 ml of pyridine and 1.1 g of NaOH in 5 ml of water. Continue H_2-bubbling for 1 h. Pour the solution together with the yellow or brown precipitate present into 200 ml of water with stirring. Continue stirring for 30 min, while yellow crystals form. Collect the yellow crystals on a sintered glass funnel. They may be contaminated with brown crystals of unreacted $[Co(DH)_2PY]_2$. Suspend the whole precipitate in 50 ml of water and stir the suspension to oxidise the remaining brown substance. Collect the yellow crystals once more using a sintered glass funnel, wash them three times with 10 ml of water and air-dry. This may be analysed for C, H and N.

Record its IR spectrum in Nujol. This compound may contain a small amount of α-isomer. To remove it, heat 1 g of the compound (in a solution of 25 ml of methanol and 100 ml of water) almost to boiling and filter hot. Collect the solid from the filter. Record the NMR spectrum of the pure β-isomer.

Preparation of methylpyridinatocobaloxime(III)

Suspend 0.5 g of β-cyanoethylpyridinatocobaloxime (crude) in 15 ml of methanol and add 1.0 ml of methyl iodide to it in a 125-ml Erlenmeyer flask. Then slowly add a solution of 0.12 g of NaOH in 30 ml of water with stirring. Keep stirring the solution. The suspended yellow crystals dissolve giving a reddish solution, and then a yellow compound precipitates again in 5–6 min. This is collected in a sintered glass funnel, washed twice with 10 ml of water and air-dried. Record its NMR spectrum (and analyse for C, H and N). A possible mechanism for this reaction is:

$$Co^{III}\text{-}CH_2CH_2CN \xrightarrow[-H_2O]{+OH^-} \begin{matrix} CH_2{=}CHCN \\ \vdots \\ Co^{I} \end{matrix} \longrightarrow Co^{III}CH_3 + I^-$$

Conversion of β-cyanoethyl to α-cyanoethyl derivative

Dissolve 0.5 g of the β-isomer in 30 ml of methanol by warming in a 125-ml Erlenmeyer flask equipped with a gas inlet side-arm. Keep H_2 bubbling through the solution and add 0.2 g of NaOH in 5 ml of water under H_2. The solution turns from the initial orange colour to deep-red. Continue H_2-bubbling (very slowly) for 1–1.5 h at room temperature. Add 5 ml of 1 mol dm^{-3} HCl aqueous solution under H_2 to neutralise the solution. Add 5 ml of water and reduce the volume of the solution to one-half to one-third on a rotary evaporator. Collect the yellow precipitate on a sintered glass funnel and air-dry (do not bother to wash with water). Record the IR spectrum (and the NMR spectrum).

Your report should include some discussion of the NMR spectra, the characteristic IR bands (e.g. CN-stretching), and a possible conversion mechanism.

Other experiments which can be performed if time allows include: preparation of methylaquocobaloxime; methylation of mercuric compounds by methylaquocobaloxime; photodecomposition of methylaquo (or pyridinato)-cobaloxime

REFERENCES

Part (A)

H. O. C. Hogenkamp (1968). *Ann. Rev. Biochem.*, **37**, 225

H. A. Baker (1972). *Ann. Rev. Biochem.*, **41**, 55–90

J. M. Wood and D. G. Brown (1972). *Structure and Bonding,* **11**, 47–105

J. M. Pratt (1972). *Inorganic Chemistry of Vitamin B_{12}*, Academic Press, New York

E. I. Ochiai (1977). *Bio-inorganic Chemistry: An Introduction,* Allyn and Bacon, Boston, Chapter 12

Part (B)

R. E. DeSimone, M. W. Penley, L. Charbonneau, S. G. Smith, J. M. Wood, H. A. O. Hill, J. M. Pratt, S. Ridsdale and R. J. P. Williams (1973). *Biochem. Biophys. Acta,* **304**, 851

V. C. W. Chu and D. W. Gruenwedel (1977). *Bio-inorganic Chemistry,* 7, 169

D. Dolphin (1971). In *Methods in Enzymology, Vol. 18, Part C,* Eds. D. B. McCormick and L. D. Wright, Academic Press, New York, 35-52. (Phenol Extraction)

Part (C)

G. N. Schrauzer and R. J. Windgassen. *J. Am. Chem. Soc.,* **89**, 1999 (1967); **88**, 3738 (1966). (Cobaloxime)

G. N. Schrauzer (1968). *Accounts Chem. Res.,* **1**, 97-103. (Cobaloxime)

Experiment 22
Catalase and its Model Compounds

INTRODUCTION

Compared to model compounds, enzymes are remarkably more efficient at catalysis. Many theories have been put forward to explain this phenomenon, but it is still somewhat of an enigma. The enzyme called catalase, which is abundant in blood, catalyses the following reaction:

$$2H_2O_2 \rightarrow 2H_2O + O_2 \uparrow \qquad (22.1)$$

Application of dilute H_2O_2 solution on a bleeding cut for disinfection produces a brisk foaming, as a result of the action of catalase in the blood.

Catalase has four Fe(III)-porphyrin centres attached to a large protein molecule, molecular weight ≑ 250 000. The most effective model compound for catalase reported so far is a Fe(III)-trien complex (trien = triethylenetetramine) which is, however, 10^3-10^4 times less effective than catalase.

This experiment demonstrates the remarkable activity of catalase compared to some simple inorganic metal complexes.

PROCEDURE

The apparatus is sketched in *Figure VIII.22.1* and comprises 1, a reaction flask, 2, a rotatable side-arm for the catalyst solution, 3, 4 and 9, stop cocks, 5, stirring bar, 6, magnetic stirrer, 7, gas burette, and 8, adjusting tube.

The procedure is as follows. Put H_2O_2 solution (10 ml) in 1, and insert 2, containing a catalyst solution. Set up the whole apparatus as shown in the figure but without water in 7 and 8. With 3 open and 4 closed, blow O_2 through the system for a few minutes, then open 4 and pour water into 7 and 8 up to the scale reading of about 0 ml, with O_2 blowing continuously. Start to stir the solution briskly using 5 and 6, and continue stirring throughout the reaction.

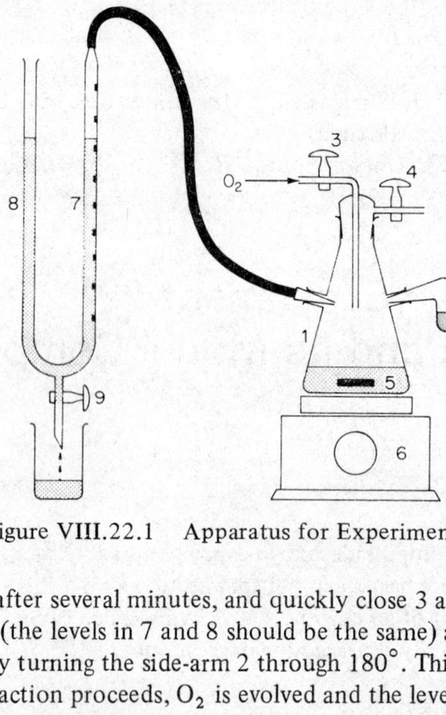

Figure VIII.22.1 Apparatus for Experiment 22

Stop the O_2 flow after several minutes, and quickly close 3 and 4. Read the level of water in 7 (the levels in 7 and 8 should be the same) and add the catalyst solution by turning the side-arm 2 through 180°. This is the start of the reaction. As the reaction proceeds, O_2 is evolved and the level in 7 will drop. Drain water by controlling 9, so that the water levels in 7 and 8 remain equal. Record the water level in 7 at 30-s intervals for about 7 min. The rapid reactions will be completed within this time.

REACTION SYSTEMS

Prepare the following solutions (of accurate concentration).

(a) H_2O_2 solution: dilute 0.80 ml of the concentrated H_2O_2 solution (about 30%) to 100 ml with a pH 9.18 buffer solution.

(b) Catalase solution: dissolve about 2.4 mg (weigh accurately) in 10 ml of water.
(b′) 1/4 concentration of (b) (> 2 ml).
(c) Trien solution, about 200 mmol dm^{-3} in water (50 ml).

The following solutions must be mixed with solution (c) immediately after preparation in a proper ratio as directed by the table below.

(d) 50 ml of about 1.0 mmol dm^{-3} aqueous solution of $Fe_2(SO_4)_3$.
(d′) 1/10 concentration of (d) (> 2 ml).
(e) 50 ml of about 2.0 mmol dm^{-3} aqueous solution of $MnSO_4$.
(e′) 1/10 concentration of (d) (> 2 ml).
(f) 50 ml of about 2.0 mmol dm^{-3} aqueous solution of $CuSO_4.5H_2O$.
(g) 50 ml of about 2.0 mmol dm^{-3} aqueous solution of $KCr(SO_4)_2.12H_2O$.

Try the following reactions. If there is insufficient time, omit reactions 7 and 8.

Reaction No.	H_2O_2 solution a (ml)	Catalyst solution (2 ml)
1	10.0	b + H_2O (1:1)
2	10.0	b′ + H_2O (1:1)
3	10.0	d′ + c (1:1)
4	10.0	e′ + c (1:1)
5	10.0	f + c (1:1)
6	10.0	c + H_2O (1:1)
7	10.0	g + c (1:1)
8	10.0	b + H_2O (1:1)
9	10.0	b′ + H_2O (1:1) + 1 mg NaCN

ESTIMATION OF THE REACTION RATES

Convert the volume change (volume of gas evolved) into the number of moles of the gas. In doing so, correct for the water vapour pressure at the reaction temperature (room temperature). Then calculate the apparent concentrations of the product; divide the moles of the gas by the total reaction volume, and express them in the unit of mol dm^{-3}. Plot the concentrations of the product against time. Measure the initial slope, which may be approximated to the apparent initial rate. In order to get the true rate, subtract the initial rate of the blank reaction (6) from the measured apparent initial rate. This blank rate is meaningful only for the reaction when the trien solution is employed. The blank rate may be neglected for the reactions 1, 2 and 8. The O_2-evolution in the reactions 1 and 8, which may be regarded as going to completion, may allow a rough estimation of the initial concentration of the H_2O_2 solution. It is better to perform a redox titration to determine the initial $[H_2O_2]$.

The reaction rate is known to follow equation 22.2:

$$\text{rate} = \frac{d[O_2]}{dt} = ½\frac{d[H_2O_2]}{dt} = ½\,k_2\,[H_2O_2]\,[\text{catalyst}] \tag{22.2}$$

Then

$$\begin{aligned}\text{initial rate} &= \text{the rate at } t = 0 \\ &= ½\,K_2\,[H_2O_2]_{\text{initial}}[\text{catalyst}]\end{aligned} \tag{22.3}$$

The rate constant k_2 may be estimated from the initial rate, since $[H_2O_2]_{\text{initial}}$ and [catalyst] are known. It may be noted here that the catalase used may not be very pure and that its activity may be as low as one tenth that of the purest catalase.

DISCUSSION

Discuss the rate data, especially with respect to the difference between metals: why is iron so effective? why is the enzyme so much more effective? You are encouraged to speculate extensively!

REFERENCES

J. H. Wang. *Accounts Chem. Res.*, **3**, 90 (1970); *J. Am. Chem. Soc.*, **77**, 4715 (1955)

R. C. Jannagin and J. H. Wang (1958). *Accounts Chem. Res.*, **80**, 786, 6477

A. S. Brill (1966). *Comprehensive Biochemistry, Vol. 14,* Eds. M. F. Florkin and E. H. Stolz, Elsevier, Amsterdam, Chapter X

B. Chance (1963). 'Investigation of Rates and Mechanisms of Reactions', Ed. A. Weissberger, *Tech. of Org. Chem.*, Vol. VIII, Part II, Interscience, 1314–1360

S. B. Brown, P. Jones and A. Suggett (1970). *Progr. Inorg. Chem.*, **13**, 159–204

D. Dolphin, A. Forman, J. Fajer and R. H. Felton (1971). *Proc. Nat. Acad. Sci., U.S.A.*, **68**, 614–618

Experiment 23
Hexokinase—Effects of Metal Ions

INTRODUCTION

There are many enzymes which require metal ions for activation. Hexokinase is an example of the so-called metal-activated enzymes. It catalyses the phosphorylation with ATP of hexoses such as glucose, mannose and fructose, and occupies an important place in the early part of carbohydrate metabolism. The reaction proceeds in accordance with the scheme below. The enzyme requires specifically Mg^{2+} for activation. In many metal-activated enzyme systems, Mg^{2+} is known to be replaced by Mn^{2+}. In this experiment, the effect of metal ions (Mg^{2+}, Mn^{2+}, Ca^{2+}, etc.) on hexokinase activity is studied.

The assay method is based on the use of an acid–base indicator (cresol red) and a buffer (glycylglycine). The amount of acid produced in the reaction is equivalent to the amount of basic buffer neutralised, which in turn is proportional to the amount of basic indicator neutralised.

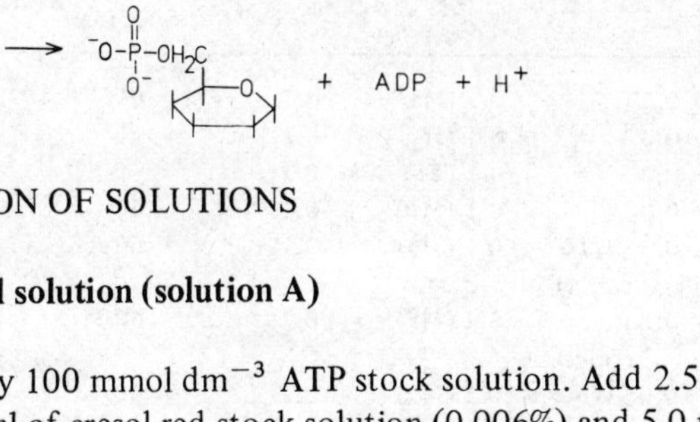

PREPARATION OF SOLUTIONS

1. ATP–cresol solution (solution A)

Thaw carefully 100 mmol dm^{-3} ATP stock solution. Add 2.5 ml of the ATP solution, 12 ml of cresol red stock solution (0.006%) and 5.0 ml of 100 mmol dm^{-3} glycylglycine stock solution to a 50-ml volumetric flask. Add to the mixture 100 mmol dm^{-3} NaOH solution with a dropping tube until the colour of the solution becomes definitely reddish-purple (do not add too much). Add water to the resultant solution and make up to 50 ml.

2. Other solutions

Available for use are 200 mmol dm^{-3} glucose stock solution (solution B), 150 mmol dm^{-3} stock solutions (solution C) of metal salts ($MgCl_2$, $CaCl_2$ and $MnSO_4$), and 1 ml of hexokinase stock solution (0.2 mg ml^{-1}, solution D).

PROCEDURE

Add 2.5 ml of solution A, 0.3 ml of solution B and 0.1 ml of solution C to a cuvette and, at reaction time zero, add 0.1 ml of solution D and mix the solution gently at 25°C. Record the reaction time counting mixing time as zero, and quickly place the cuvette in a spectrophotometer. Record the absorbance at 570 nm at 10-s intervals or record it automatically on a chart. The reactions recommended are listed in *Table VIII.23.1.* From Reaction Nos. 8, 8′ and 8″, the absorbance change due to the increase in $[H^+]$ is determined. Determine the initial rate (in mol dm^{-3} s^{-1}) from the initial slope of the absorbance versus time curve. Using the data of Reaction Nos. 1, 5, 6, and 7, the Michaelis–Menten parameters K_m and V_m may be estimated.

Table 23.1
Solutions for Investigating Hexokinase Reactions

Reaction No.	*A*	*B*	*C*	*D*	Chart speed (mm min^{-1})	Reaction time to be followed (min)
1	2.5	0.3	0.1 (Mg)	0.1	60	2
2	2.5	0.3	0.1 (H_2O)	0.1	5	10
3	2.5	0.3	0.1 (Ca)	0.1	5	10
4	2.5	0.3	0.1 (Mn)	0.1	60	2
5	2.5	0.3 (1/10 conc.)	0.1 (Mg)	0.1	60	2
6	2.5	0.3 (1/20 conc.)	0.1 (Mg)	0.1	60	2
7	2.5	0.3 (1/30 conc.)	0.1 (Mg)	0.1	60	2
8	2.5	0.3	0.1 (Mg)	0.1 (10 mmol dm^{-3} HCl)		
8′		8 + 0.1 ml (10 mmol dm^{-3} HCl)				
8″		8′ + 0.1 ml (10 mmol dm^{-3} HCl)				

DISCUSSION

Discuss the effect of metal ions on hexokinase activity.

REFERENCES

R. A. Darrow and S. P. Colowick (1962). In *Methods in Enzymology, Vol. 5,* Eds. S. P. Colowick and N. O. Kaplan, Academic Press, New York.
R. K. Crane, (1962). In *The Enzymes, Vol. 6,* Eds. P. D. Boyer, H. Lardy and K. Myrback, Academic Press, New York
R. J. P. Williams (1970). *Quart. Rev.*, **24**, 331

Experiment 24
The Effects of Metal Ions on Polynucleotides (DNA)

INTRODUCTION

It has long been known that native preparations of nucleic acids, especially RNA, contain significant levels of metal ions such as magnesium, calcium, aluminium, chromium, manganese, iron, nickel and zinc. Some of these metal ions, particularly magnesium(II), are considered to be important in stabilising the conformation of the nucleic acids DNA and RNA.

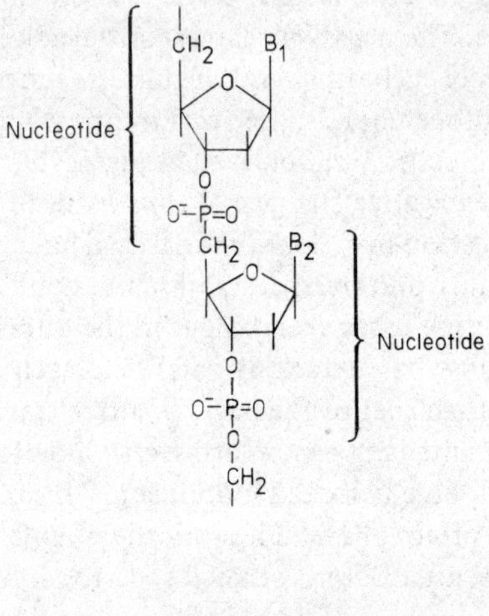

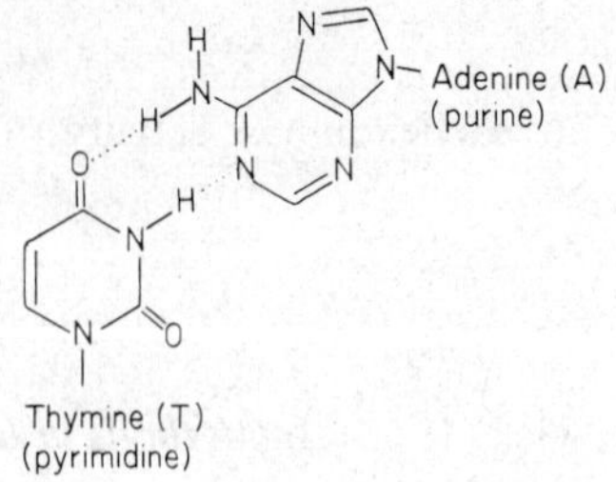

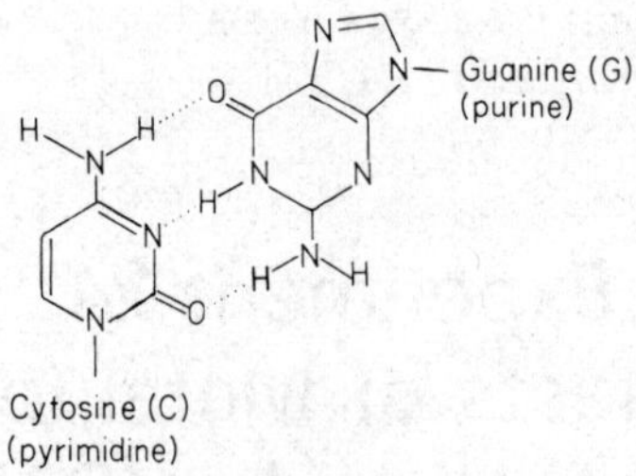

The nucleic acid consists of nucleotide units; a nucleotide is made of a pentose (ribose in RNA and deoxyribose in DNA), a phosphate group which bridges two nucleotide units, and a base (*see* Scheme IX). The base is either a purine or pyrimidine; the bases in DNA are shown in the schemes above. As illustrated, A and T pair up owing to the two hydrogen bonds, and G and C pair up likewise. This is the very basis of the double helix structure of DNA. The molecular planes of the bases stack on top of each other in the double helix structure of nucleic acid. The potential coordinating (to metal ions) atoms in DNA are the oxygen of the phosphate group, the carbonyl oxygens and the amine nitrogens of the bases. The phosphate groups on the backbone of the polynucleotides, being negatively charged, destabilise the double helix structure of a nucleic acid. The negative charges on the adjoining two strands of the polynucleotides repel each other and, if this effect overcomes the cohesive effect of the hydrogen bonds, the two strands would separate and form single-strand random coils. A cation such as Mg(II) binds to the phosphate groups and neutralises the negative charge and thus tends to stabilise the double helix structure. Another cation may coordinate to the base portion, break up some of the hydrogen bonds, and thus, destabilise the double helix structure.

The pyrimidine and purine bases absorb light in the ultraviolet region. The absorption intensity of those bases stacking parallel to each other in the double helix structure is smaller than that of the bases in a free state, as in a random coil. Thus, the absorption intensity can be utilised to monitor the structure transition of DNA from double helix to random coil. This is the basis of this experiment, in which the effect of metal ions on the double helix–random coil transition temperature of a DNA is studied spectrophotometrically.

PROCEDURE

DNA solution

Some DNA preparations are commercially available. For example, a DNA from calf thymus is available from Sigma Chemical Co. This particular sample seems to contain a substantial quantity of metals, very likely Mg(II). Approximately 1×10^{-4} mol dm^{-3} (in phosphate) solution is appropriate for the experiment. Since the average molecular weight of nucleotides is 300, a 1×10^{-4} mol dm^{-3} solution requires 3 mg of DNA for 100 ml. Prepare a 100-ml 1×10^{-4} mol dm^{-3} (in nucleotide) DNA solution dissolved in 1×10^{-2} mol dm^{-3} $NaNO_3$ solution.

Metal salt solutions

Prepare 100 ml of 2×10^{-4} mol dm^{-3} aqueous solution each of either the nitrate or sulphate of copper, magnesium and zinc.

Experimental setup

This consists of a spectrophotometer capable of working in the ultraviolet region and a heating system. A Turner spectrophotometer (or similar) will be appropriate. The quartz cell should be water-jacketed unless the cell compartment is equipped with a heating device. The temperature of the cell (or cell compartment) must be read. If this is not feasible, the temperature of the circulator-reservoir may be read; but some correction should be made for the difference between the temperature read and the temperature of the cell compartment. The setup is to be heated as high as 95°C and so all tubing, etc. must be of materials which can stand this temperature.

The cell must be stoppered securely otherwise solvent will evaporate at higher temperatures.

Measurements

Prepare DNA-metal ion solutions by mixing the DNA solution and a metal salt solution in 1:1 ratio, as follows: (a) DNA-H_2O, (b) DNA-Mg(II), (c) DNA-Cu(II), and (d) DNA-Zn(II).

Add one of the solutions to a cell to about 2/3 full, stopper the cell and insert into the cell compartment. Set the wavelength to 260 nm and record the absorbance and temperature. Raise the temperature at a rate of about 1° per minute. Record the absorbance and the temperature as the temperature rises. The absorbance will increase sharply around the double helix–random coil transition temperature and will then stabilise. Follow this flat portion for another 5-10° and then start to lower the temperature. Again record the absorbance versus temperature. Plot the absorbance versus temperature and determine the transition temperature. Repeat the same procedure for all the other solutions.

DISCUSSION

Compare the transition temperatures and discuss possible modes of interaction of the metal ions with DNA.

REFERENCES

G. L. Eichhorn (1962). *Nature,* **194**, 474

G. L. Eichhorn, N. A. Berger, J. J. Butzow, P. Clark, J. M. Rifkind, Y. A. Shin and E. Tarien (1971). *Adv. Chem. Ser.,* **100**, 135

E. -I. Ochiai (1977). *Bio-inorganic Chemistry: An Introduction,* Allyn and Bacon, Boston, Chapter 14

Experiment 25 Computer Simulations of Complexing Systems *in vivo*

INTRODUCTION

Sillén has used computer models to simulate the equilibria involved between air, sea and sediment[1] and Perrin[2] has published models of blood plasma that have recently been extended to 9000 species models[3–5]. In principle all bio-fluids contain metal ions present in four forms:

(i) Those firmly complexed within metalloproteins (e.g. the copper in ceruloplasmin, the zinc in α_2-macroglobulin and the iron in haemoglobin). These metal ions are inert and cannot be exchanged without destroying the whole protein.
(ii) Those metal ions more loosely complexed to other proteins (e.g. copper and zinc reversibly complexed to serum albumin and the ferric-transferrin equilibrium).
(iii) Those associated with the many low-molecular-weight ligands present.
(iv) Those present as aquated ions (in practice because of the high preponderance of ligands, fraction (iv) is usually very low).

Portions (ii), (iii) and (iv) are in labile equilibrium so that, for example, adding additional low-molecular-weight ligands can pull metal from (ii) to (iii) and (iv), and similarly adding metal causes an across-the-board redistribution involving (ii), (iii) and (iv).

It is important to realise that it is only the low-molecular-weight type of metal complexes which are absorbed from the intestine, which are transported between blood plasma and tissues, which are excreted in bile and urine, and which are incorporated into metalloenzymes. Similarly, if one wishes to promote the renal excretion of a metal one must first mobilise it as low-molecular-weight complexes from labile protein (type (ii)) complexes and then filter off these complexes in the kidneys. Thus it is imperative to have an accurate knowledge of the concentrations of these low-molecular-weight complexes in all biofluids under all conditions. Unfortunately, although the total of each metal present as (i) + (ii) + (iii) + (iv) can be easily measured by atomic absorption spectrophotometry and, using ultrafiltration techniques, (iii) + (iv) can be established, the values of individual complex concentrations for low-molecular-weight complexes are orders of magnitude below those analysable by any known analytical method. Computer simulation is the only means available for estimating these concentrations.

In principle each metal–ligand and proton–ligand reaction possible *in vivo* is characterised *in vitro* in terms of its formation constants. Further metal ion hydrolysis and complex hydrolysis constants are also determined. Next the total concentration of metal and ligand present in low-molecular-weight form is established and used as input. The computer programs then combine all these equilibrium reactions and compute the concentrations present for *in vivo* conditions. Currently such models are capable of handling more than 9000 species.

When exogenous ligands are added to a biofluid they often promote the removal of metal from protein into low-molecular-weight form, i.e. they mobilise the metal ion. Further we are often faced with the problem of different patients having different plasma drug levels of different drugs and so a plasma mobilising index (P.M.I.) is defined as the factor by which, for a given concentration of chelating agent, the total low-molecular-weight fraction of metal ion is increased. P.M.I. = (total low-molecular-weight fraction in the presence of drug)/(total low-molecular-weight species in normal plasma).

PROCEDURES

The programs often used in computer simulation, COMICS[6], or its graph plotter version, COMPLOT[7], ECCLES[3] and HALTAFALL[8] are readily available and you should enquire whether they are loaded on to the computer that serves your laboratory. However, even whilst awaiting such facilities, one can gain valuable experience concerning the best choice of data to be used as input and the interpretation of the computed output.

Table VIII.25.1.
Percentage Distribution of the Metal Ions Ca^{2+}, Cu^{2+}, Fe^{3+} Pb^{2+}, Mg^{2+}, Mn^{2+}, and Zn^{2+} amongst Low-molecular-weight Ligands in Human Blood Plasma as Computed using the ECCLES Program[a]

Complex	Charge	Percentage of the total metal in the low-molecular-weight fraction at pH = 7.4
(a) Calcium (II)		
Protonated carbonate	+1	9
Citrate	−1	4
Lactate	+1	3
Phosphate	−1	3
Carbonate	0	2
(b) Copper (II)		
Cystinate histidinate	−1	21
Protonated cystinate histidinate	0	17
Bis(histidinate)	0	11
Histidinate threoninate	0	8
Histidinate valinate	0	5
Protonated histidinate lysinate	+1	5
Alanate histidinate	0	4
Histidinate serinate	0	4
(c) Iron (III)		
Citrate hydroxide	−1	99
(d) Lead (II)		
Cysteinate	0	80
Citrate cysteinate	−3	7
Protonated cystinate	+1	5
(e) Magnesium(II)		
Protonated carbonate	+1	6
Citrate	−1	5
(f) Manganese(II)		
Protonated carbonate	+1	24
Citrate	−1	10
(g) Zinc(II)		
$Zn.(cysteinate)_2$	−2	40
Zn. cysteinate.histidinate	−1	24

[a]For a complete listing of percentages see References 3–5.

Choice of input data

The formation constants (β) and total concentrations used as input data are critical. Such constants published until a few years ago are available as compendiums[9,10]. More recent values can be traced through *Chemical Abstracts* or measured directly as described in Experiment 11. To familiarise yourself with these sources select one or two metal ligand systems found to be important in reference 3 and search the literature for values of β reported. Do you think the authors of reference 3 have made the correct choice of β values? Next, calculate a theoretical value for the ternary formation constant based upon the β values for its parent binaries (as described in chapter 6 of reference 11).

The total concentrations of ligands and metals are also available from the literature. Particularly useful compilations are given in references 12 and 13. However, it is also reasonably easy to measure total metal concentrations for actual biological samples as described in Experiments 3-7.

Understanding the computed output

A typical percentage of total metal output is given for blood plasma in *Table VIII.25.1* and a set of log(P.M.I.) curves for administered therapeuticals are illustrated in *Figure VIII.25.1.* Note the overall charges on the complexes and divide the list into those which are charged and therefore will either remain in plasma or be excreted through the kidneys (renal filtration is a type of desalination process as far as low-molecular-weight complexes are concerned), and those that can possibly penetrate lipid-protein cell membranes into cells (the neutral complexes). Clearly, if one wishes to encourage cell penetration or urinary excretion one adds therapeuticals which enhance the concentration gradients of desired complexes.

In the light of these remarks and with particular reference to reference 5 explain why plumbism treatment involves initially EDTA followed by long-term D-penicillamine and note the dangers of premature D-penicillamine therapy.

Turning to the log(P.M.I.) curves, show how D-penicillamine therapy promoted zinc (a) mobilisation and (b) excretion. Which are the better agents for copper mobilisation, 10^{-8} mol dm^{-3} triethylenetetramine or 10^{-4} mol dm^{-3} EDTA? Similarly, which mobilises zinc better, 10^{-6} mol dm^{-3} DTPA or 10^{-5} mol dm^{-3} triethylenetetramine?

COMPUTATIONAL EXERCISES

(A) Read the paper entitled 'Analytical and Computer Simulation Studies of a Colloidal Bismuth Citrate System Used as an Ulcer Treatment' in reference 14 and set up some of the models described using the HALTAFALL program

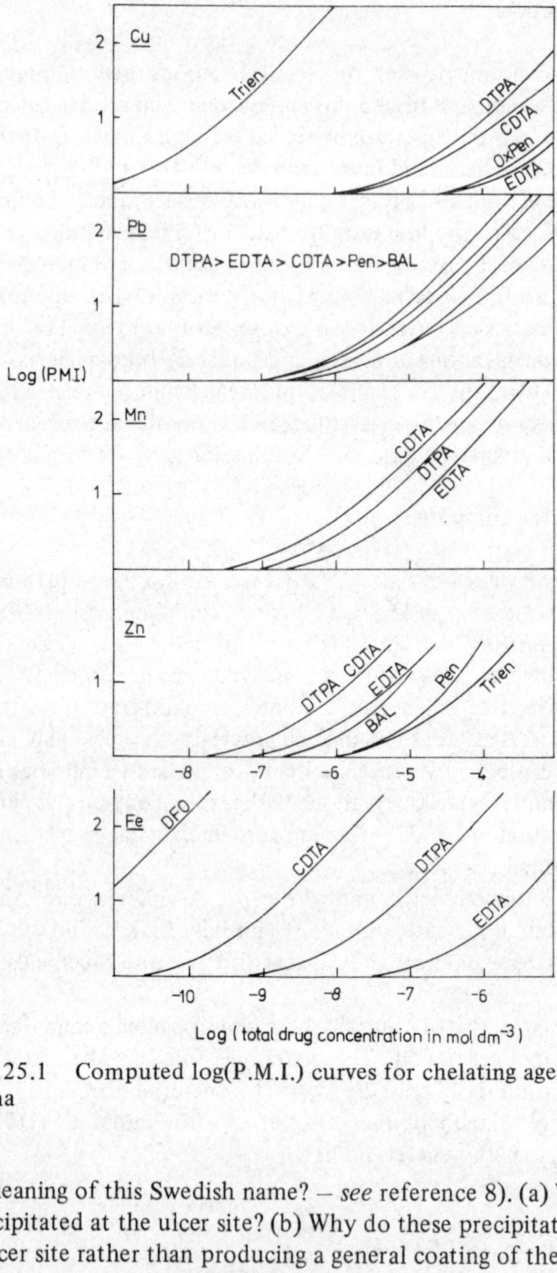

Figure VIII.25.1 Computed log(P.M.I.) curves for chelating agents added to blood plasma

(what is the meaning of this Swedish name? – *see* reference 8). (a) What are the chemicals precipitated at the ulcer site? (b) Why do these precipitates selectively form at the ulcer site rather than producing a general coating of the stomach lining? (c) Compare the advantages of producing improved formulations from computer simulation compared with empirical screening in humans.

(B) Taking the constants determined in Experiment 11, use the COMPLOT[7,6] program to calculate *in vitro* concentrations of complexes present plotted versus pH (over the range 1–13) for total metal ion = 10^{-3} mol dm^{-3} and ligand = 3×10^{-3} mol dm^{-3}. Assuming that small errors may have been made in determining these constants, alter their values by a small amount and repeat the COMPLOT computation. Is the model sensitive to small changes in log β?

REFERENCES

1 L. G. Sillén (1967). *Chem. Brit.*, 291

2 D. D. Perrin and R. P. Agarwal (1973). *Metal Ions in Biological Systems, Vol. 2,* Ed. H. Sigel, Marcel Dekker, New York, 167

3 P. M. May, P. W. Linder and D. R. Williams (1977). *J. Chem. Soc. Dalton,* 588, (1978) 1433

4 P. M. May, P. W. Linder and D. R. Williams (1976). *Experientia,* **32**, 1492

5 P. M. May and D. R. Williams (1977). *F. E. B. S. Letters,* **78**, 134

6 D. D. Perrin and I. G. Sayce (1967). *Talanta,* **14**, 833

7 A. C. Baxter and D. R. Williams (1974). *J. Chem. Soc. Dalton,* 1117

8 N. Ingri, W. Kakotowicz, L. G. Sillén and B. Warnqvist (1967). *Talanta,* **14**, 1261

9 *Stability Constants,* The Chemical Society, London, Spec. Publs. Nos. 17 and 25 (1964 and 1971)

10 *Critical Stability Constants,* Eds. A. E. Martell and R. M. Smith, Plenum, New York (1974)

11 D. R. Williams (Ed.), *An Introduction to Bio-inorganic Chemistry,* C. C. Thomas, Illinois (1976)

12 *The Biochemists' Handbook,* Ed. C. Long, SPON, London (1961)

13 K. Diem and C. Lentnes (Eds.), *Documenta Geigy Scientific Tables,* Geigy Pharmaceuticals, Macclesfield, 7th edn. (1975)

14 D. R. Williams (1977). *J. Inorg. Nucl. Chem.,* **39**, 711

Index